William Jones Rhees, James Smithson, Walter Rogers Johnson, John Robin McDaniel Irby

The scientific writings of James Smithson

William Jones Rhees, James Smithson, Walter Rogers Johnson, John Robin McDaniel Irby

The scientific writings of James Smithson

ISBN/EAN: 9783337414337

Printed in Europe, USA, Canada, Australia, Japan

Cover: Foto ©berggeist007 / pixelio.de

More available books at **www.hansebooks.com**

SMITHSONIAN MISCELLANEOUS COLLECTIONS.
——— 327 ———

THE SCIENTIFIC WRITINGS

OF

JAMES SMITHSON.

EDITED BY

WILLIAM J. RHEES.

WASHINGTON:
PUBLISHED BY THE SMITHSONIAN INSTITUTION.
1879.

ADVERTISEMENT.

The scientific writings of James Smithson, the distinguished founder of the Smithsonian Institution, have been collected and are published in the present volume, in accordance with the instructions of the Board of Regents. These memoirs were originally contributed to the "Transactions of the Royal Society of London," of which Smithson was a member, between the years of 1791 and 1817, and to Thomson's "Annals of Philosophy," between 1819 and 1825. They are twenty-seven in number, and embrace a wide range of research, from the origin of the earth, the nature of the colors of vegetables and insects, the analysis of minerals and chemicals, to an improved method of constructing lamps or of making coffee. Some of these papers were translated into French by the author and others, and published in the "Journal de Physique, de Chimie, et d'Histoire Naturelle, etc."

These writings of Smithson prove conclusively his scientific character and his claim to distinction as a contributor to knowledge.

Among the personal effects of the founder of the Institution were several hundred manuscripts, besides a large collection of scraps and notes on a great diversity of subjects, including history, the arts, language, rural economy, construction of buildings, &c., which unfortunately were destroyed by the fire at the Smithsonian building in 1865. It is probable that Smithson also contributed articles to other scientific and literary journals than those mentioned, but none have been found, though the leading English periodicals of the day have been carefully examined for the purpose.

Appended to the writings of Smithson is a review of their scientific character by Professor Walter R. Johnson, communicated to the National Institute, of Washington, in 1844; and one by J. R. McD. Irby, prepared for the Institution in September, 1878. The material for this work has been collected and prepared for publication by Mr. Wm. J. Rhees, Chief Clerk of the Institution.

SPENCER F. BAIRD,
Secretary Smithsonian Institution.

WASHINGTON, D. C., *October*, 1879.

CONTENTS.

II.—Reviews.

AN ACCOUNT OF SOME CHEMICAL EXPERIMENTS ON TABASHEER.

From the Philosophical Transactions of the Royal Society of London. Vol. LXXXI, for the year 1791, Part 2, p. 368.—Read July 7, 1791.

The Tabasheer employed in these experiments was that which Dr. Russell laid before the Society, as specimens of this substance, the evening his Paper upon the subject was read.*

There were seven parcels.

No. 1 consisted of Tabasheer extracted from the bamboo by Dr. Russell himself.

No. 2 had been partly taken from the reed in Dr. Russel's presence, and partly brought to him at different times by a person who worked in bamboos.

No. 3 was the Tabasheer from Hydrabad; the finest kind of this substance to be bought.

Nos. 4, 5, and 6 all came from Masulapatam, where they are sold at a very low price. These three kinds have been thought to be artificial compositions in imitation of the true Tabasheer, and to be made of calcined bones.

No. 7 had no account affixed to it.

The Tabasheer from Hydrabad being in the greatest quantity, and appearing the most homogeneous and pure, the experiments were begun, and principally made, with it.

Hydrabad Tabasheer. (No. 3.)

§ I. (A) This, in its general appearance, very much resembled fragments of that variety of calcedony which is known to mineralogists by the name of *Cacholong*. Some pieces were quite opaque, and absolutely white; but others

---

* See Phil. Trans. Vol. LXXX, p. 283.

possessed a small degree of transparency, and had a bluish cast. The latter, held before a lighted candle, appeared very pellucid, and of a flame colour.

The pieces were of various sizes; the largest of them did not exceed two or three-tenths of an inch cubic. Their shape was quite irregular; some of them bore impressions of the inner part of the bamboo against which they were formed.

(B) This Tabasheer could not be broken by pressure between the fingers; but by the teeth it was easily reduced to powder. On first chewing it felt gritty, but soon ground to impalpable particles.

(C) Applied to the tongue, it adhered to it by capillary attraction.

(D) It had a disagreeable earthy taste, something like that of magnesia.

(E) No light was produced either by cutting it with a knife, or by rubbing two pieces of it together, in the dark; but a bit of this substance, being laid on a hot iron, soon appeared surrounded with a feeble luminous *auréole*. By being made red hot, it was deprived of this property of shining when gently heated; but recovered it again, on being kept for two months.

(F) Examined with the microscope, it did not appear different from what it does to the naked eye.

(G) A quantity of this Tabasheer which weighed 75.7 gr. in air, weighed only 41.1 gr. in distilled water whose temperature was 52.5 F. which makes its specific gravity to be very nearly = 2.188.

Mr. CAVENDISH, having tried this same parcel when become again quite dry, found its specific gravity to be = 2.169.

Treated with water.

§ II. (A) This Tabasheer, put into water, emitted a number of bubbles of air; the white opaque bits became transparent in a small degree only, but the bluish ones nearly as much so as glass. In this state the different colour produced by reflected and by transmitted light was very sensible.

(B) Four bits of this substance, weighing together, while dry and opaque, 4.1 gr., were put into distilled water, and let become transparent; being then taken out, and the unabsorbed water hastily wiped from their surface, they were again weighed, and were found to equal 8.2 gr.

In the experiment § I. (G), 75.7 gr. of this substance absorbed 69.5 gr. of distilled water.

(C) Four bits of Tabasheer, weighing together 3.2 gr. were boiled for 30' in half an ounce of distilled water in a Florence flask, which had been previously rinced with some of the same fluid. This water, when become cold, did not shew any change on the admixture of vitriolic acid, of acid of sugar, nor of solutions of nitre of silver, or of crystals of soda; yet, on its evaporation, it left a white film on the glass, which could not be got off by washing in cold water, nor by hot marine acid; but which was discharged by warm caustic vegetable alkali, and by long ebullition in water.

Upon these bits of Tabasheer, another half ounce of distilled water was poured, and again boiled for about half an hour. This water also on evaporation left a white film on the glass vessel similar to the above. The pieces of Tabasheer having been dried, by exposure to the air for some days in a warm room, were found to have lost one-tenth of a grain of their weight.

To ascertain whether the whole of a piece of Tabasheer could be dissolved by boiling in water, a little bit of this substance, weighing three-tenths of a grain, was boiled in 36 ounces of soft water for near five hours consecutively; but being afterwards dried and weighed, it was not diminished in quantity, nor was it deprived of its taste.

With vegetable colours.

§ III. Some Tabasheer, reduced to fine powder, was boiled for a considerable time in infusions of turnsole, of logwood,

and of dried red cabbage, but produced not the least change
in any one of them.

At the fire.

§ IV. (A) A piece of this Tabasheer, thrown into a red
hot crucible, did not burn or grow black. Kept red hot for
some time, it underwent no visible change; but when cold,
it was harder, and had entirely lost its taste. Put into water
it grew transparent, just as it would have done, had it not
been ignited.

(B) 6.4 gr. of this substance, made red hot in a crucible,
were found, upon being weighed as soon as cold, to have
lost two-tenths of a grain. This loss appears to have arisen
merely from the expulsion of interposed moisture; for these
heated pieces, on being exposed to the air for some days,
recovered exactly their former weight.

(C) A bit of this substance was put into an earthen cru-.
cible, surrounded with sand, and kept red hot for some time;
when cold, it was still white both exteriorly and interiorly.

(D) Thrown into some melted red hot nitre, this substance
did not produce any deflagration, or seem to suffer any alteration.

(E) A bit exposed on charcoal to the flame of the blow-
pipe did not decrepitate or change colour; when first heated
it diffused a pleasant smell; then contracted very considerably in bulk, and became transparent; but on continuing
the heat it again grew white and opaque, but seemed not to
shew any inclination to melt *per se*. Possibly, however, it
may suffer such a semi-fusion, or softening of the whole
mass, as takes place in clay when exposed to an intense
heat; for when the bit used happened to have cracks, it
separated during its contraction, at these cracks, and the
parts receded from each other without falling asunder.

If, while the bit of Tabasheer was exposed to the flame,
any of the ashes of the coal fell upon it, it instantly melted,
and small very fluid bubbles were produced. That the
opacity which this substance acquires on continuing to heat

it after it is become transparent, is not owing to the fusion of its surface by means of some of the ashes of the charcoal settling upon it unobserved, appeared by its undergoing the same change when fixed to the end of a glass tube, in the method of M. DE SAUSSURE.*

With acids.

§ V. (A) A piece of Tabasheer, weighing 1.2 gr. was first let satiate itself with distilled water; its surface being then wiped dry, it was put into a matrass with some pure white marine acid, whose specific gravity was 1.13. No effervesence arose on its immersion into the acid; nor did this menstruum, even by ebullition, seem to have any action upon it, or itself receive any colour. The acid being evaporated left only some dark coloured spots on the glass. These spots were dissolved by distilled water. No precipitation was produced in this water by vitriolic acid, or by a solution of crystals of soda. The bit of Tabasheer washed with water, and made red hot, had not sustained any loss of weight.

The pores of the mass of Tabasheer were filled with water before it was put into the acid, to expel the common air contained in them, and which would have made it impossible to ascertain with accuracy whether any effervescence was produced on its first contact with the menstruum.

(B) Another portion of Tabasheer, weighing 10.2 gr. was boiled in some of the same marine acid. Not the least precipitate was produced on saturating this acid with solution of mild soda. This Tabasheer also, after having been boiled in water, and dried by exposure for some days to the air, was still of its former weight.

§ VI. This substance seemed in like manner to resist the action of pure white nitrous acid boiled upon it.

§ VII. (A) A bit of Tabasheer weighing 0.6 gr. was digested in some strong white vitriolic acid, which had been

---

* Journal de Physique, Tom. XXVI, p. 409.

made perfectly pure by distillation. It did not seem by this
treatment to suffer any change, and after having been freed
from all adhering vitriolic acid by boiling in water, it had
not undergone any alteration either in its weight or properties. The vitriolic acid afforded no precipitate on being
saturated with soda.

(B) Two grains of Tabasheer reduced to fine powder were
made into a paste with some of this same vitriolic acid, and
this mixture was heated till nearly dry; it was then digested
in distilled water. This water, being filtered, tasted slightly
acid, did not produce the least turbidness with solution of
soda, and some of it, evaporated, left only a faint black
stain on the glass, produced doubtless by the action of the
vitriolic acid on a little vegetable matter, which it had received either from the Tabasheer, or from the paper. The
undissolved matter collected, washed, and dried, weighed
1.9 gr.

§ VIII. 2 gr. of Tabasheer, reduced to fine powder, were
long digested in a considerable quantity of liquid acid of
sugar. The taste of the liquor was not altered; and being
saturated with a solution of crystals of soda in distilled
water, it did not afford any precipitate. The Tabasheer having been freed from all adhering acid, by very careful ablution with distilled water, and let dry in the air, was totally
unchanged in its appearance, and weighed 1.98 gr. This
Tabasheer being gradually heated till red hot, did not
become in the least black, or lose much of its weight, a
proof that no acid of sugar had fixed in it.

With liquid alkalies.

§ IX. (A) Some liquid caustic vegetable alkali being
heated in a phial, Tabasheer was added to it, which dissolved very readily, and in considerable quantity. When
the alkali would not take up any more, it was set by to cool,
but was not found next morning to have crystallized, or undergone any change, though it had become very concen-

trated, during the boiling, by the evaporation of much of the water.

(B) This solution had an alkaline taste, but seemingly with little, if any, causticity.

(C) A drop of it changed to green a watery tincture of dried red cabbage.

(D) Some of this solution was exposed in a shallow glass to spontaneous evaporation in a warm room. At the end of a day or two it was converted into a firm, milky jelly. After a few days more, this jelly was become whiter, more opaque, and had dried and cracked into several pieces, and finally it became quite dry, and curled up and separated from the glass.

The same change took place when the solution had been diluted with several times its bulk of distilled water, only the jelly was much thinner, and dried into a white powder.

Some of this solution, kept for many weeks in a bottle closely stopped, did not become a jelly, or undergo any change.

(E) A small quantity of this solution was let fall into a proportionably large quantity of spirit of wine, whose specific gravity was .838. The mixture immediately became turbid, and, on standing, a dense fluid settled to the bottom, and which, when the bottle was hastily inverted, fell through the spirit of wine in round drops, like a ponderous oil.

The supernatant spirit of wine being carefully decanted off, some distilled water was added to this thick fluid, by which it was wholly dissolved. This solution, exposed to the air, shewed phænomena exactly similar to those of the undiluted solution (D).

The decanted spirit being also left exposed to the air in a shallow glass vessel, did not, after many days, either deposit a sensible quantity of precipitate, or become gelatinous; but having evaporated nearly away, left a few drops of a liquor which made infusion of red cabbage green; and, on the addition of some pure marine acid, effervesced violently. No precipitate fell during this saturation with the acid; nor did the mixture on standing become a jelly; and on the total evaporation of the fluid part, a small quantity of muriate of tartar only remained. The spirit of wine seems, therefore, to have dissolved merely a portion of superabundant alkali present in the mixture, but none of that united with Tabasheer.

(F) To different portions of this solution were added some pure marine acid, some pure white vitriolic acid, and some distilled vinegar, each in excess. These acids at first produced neither heat, effervescence, any precipitate, or the least sensible effect, except the vitriolic acid, which threw down a very small quantity of a white matter; but, after standing some days, these mixtures changed into jellies so firm, that the glasses containing them were inverted without their falling out.

This change into jelly equally took place whether the mixtures were kept in open or closed vessels, were exposed to the light or secluded from it; nor did it seem to be much promoted by boiling the mixtures.

(G) Some solution of mild volatile alkali in distilled water, being added to some of this solution, seemed at the first instant of mixture to have no effect upon it; but in the space of a second or two it occasioned a copious white precipitate.

(H) The flakes remaining on the glasses at (D) and (E) put into marine acid raised a slight effervescence, but did not dissolve. These flakes when taken out of the acid, and well washed, were found, like the original Tabasheer, to be white and opaque when dry; but to become transparent when moistened, and then to shew the blue and flame colour, § II. (A).

(I) The jellies (F), diluted with water, and collected on a filter, appeared to be the Tabasheer unchanged.

§ X. A bit of Tabasheer, weighing two-tenths of a grain, was boiled in 127 gr. of strong caustic volatile alkali for a considerable time; but after being made red hot, it had not sustained the least diminution of weight.

§ XI. (A) 27 gr. of Tabasheer reduced to fine powder, were put into an open tin vessel with 100 gr. of crystals of soda, and some distilled water, and this mixture was made boil for three hours. The clear liquor was then poured off, and the Tabasheer was digested in some pure marine acid; after some time this acid was decanted, and the Tabasheer washed with distilled water, which was then added to the acid.

(B) This Tabasheer was put back into the alkaline solution, which seemed not impaired by the foregoing process, and again boiled for a considerable time. The liquor was then poured from it while hot, and the Tabasheer edulcorated with some cold distilled water, which was afterwards mixed with this hot solution, in which it instantly caused a precipitation. On heating the mixture it became clear again; but as it cooled it changed wholly into a thin jelly; but in the course of a few days, it separated into two portions, the jelly settling in a denser state to the bottom of the vessel, leaving a limpid liquor over it.

(C) The Tabasheer remaining (B) was boiled in pure marine acid; the acid was then poured off, and the Tabasheer edulcorated with some distilled water, which was afterwards mixed with the acid.

(D) The remaining Tabasheer collected, washed, and dried, weighed 24 gr. and seemed not to be altered.

(E) The acid liquors (A and C) were mixed together, and saturated with soda, but afforded no precipitate.

(F) The alkaline mixture (B) was poured upon a filter, the clear liquor came through, leaving the jelly on the paper. Some of this clear liquor, exposed to the air in a saucer, at the end of some days deposited a small quantity of a gelatinous matter; after some days more, the whole fluid part exhaled, and the saucer became covered with regular crystals of soda, which afforded no precipitate during their solution in vitriolic acid. What had appeared like a jelly while moist, assumed, on drying, the form of a white powder.

This powder was insoluble in vitriolic acid, and seemed still to be Tabasheer.

Some of this clear liquor, mixed with marine acid, effervesced; did not afford any precipitate; but, on standing some days, the mixture became slightly gelatinous.

(G) Some of the thick jelly remaining on the filter, being boiled in water and in marine acid, appeared insoluble in both, and seemed to agree entirely with the above powder (F).

With dry alkalies.

§ XII. (A) Tabasheer melted on the charcoal at the blowpipe with soda, with considerable effervescence. When the proportion of alkali was large, the Tabasheer quickly dissolved, and the whole spread on the coal, soaked into it, and vanished; but, by adding the alkali to the bit of Tabasheer in exceedingly small quantities at a time, this substance was converted into a pearl of clear colourless glass.

(B) 5 gr. of Tabasheer, reduced to fine powder, were melted in a platina crucible with 100 gr. of crystals of soda. The mass obtained was white and opaque, and weighed 40.2 gr. Put into an ounce of distilled water, it wholly dissolved. An excess of marine acid let fall into this solution produced an effervescence, and changed it into a jelly. This mixture was stirred about, and then thrown upon a filter. The jelly left on the paper did not dissolve in marine acid by ebullition; collected, washed with distilled water, and dried, it weighed 4.5 gr. and seemed to be the Tabasheer unaltered.

The liquor which had come through being saturated with mineral alkali yielded only a very small quantity of a red precipitate, which was the colouring matter of the pink blotting paper through which it had been passed.

(C) 10 gr. of Tabasheer, reduced to powder, were mixed with an equal weight of soda, deprived of its water of crystallization by heat. This mixture was put into a platina crucible, and exposed to a strong fire for 15'. It was then found converted into a transparent glass of a slight yellow colour. This glass was broken into pieces, and boiled in marine acid. No effervescence appeared; but the glass was dissolved into a jelly. This jelly, collected on a filter, well washed and dried, weighed 7.7 gr.

The acid liquor which came through, on saturation with soda, afforded not the least precipitate; but, after standing a day or two, it changed into a thin jelly. This collected on a filter was washed with distilled water, and then boiled in marine acid, but did not dissolve. Being again edulcorated, and made red hot, it weighed 1.6 gr. The filtered liquor (B) would in all probability have changed similarly to a jelly, had it been kept. These precipitates were analogous to those § IX. (I).

(D) An equal weight of vegetable alkali and Tabasheer were melted together in the platina crucible. The glass produced was transparent; but it had a fiery taste, and soon attracted the moisture of the air, and dissolved into a thick liquor. But two parts of vegetable alkali, with three of Tabasheer, yielded a transparent glass, which was permanent.

Treated with other fluxes.

§ XIII. (A) A fragment of Tabasheer put into glass of borax, and urged at the blow-pipe, contracted very considerably in size, the same as when heated *per se;* after which it continued turning about in the flux, dissolving with great difficulty and very slowly. When the solution was effected, the saline pearl remained perfectly clear and colourless.

(B) With phosphoric ammoniac (made by saturating the acid obtained by the slow combustion of phosphorus with caustic volatile alkali) the Tabasheer very readily melted on the charcoal at the blow-pipe, with effervescence, into a white frothy bead.

(C) Fused, by the same means, on a plate of platina, with the vitriols of tartar and soda, it appeared entirely to resist their action; the little particles employed continuing to revolve in the fluid globules without sustaining any sensible diminution of size, and the saline beads on cooling assumed their usual opacity.

(D) A bit of Tabasheer was laid on a plate of silver, and a little litharge was put over it, and then melted with the blow-pipe. It immediately acted on the Tabasheer, and covered it with a white glassy glazing. By the addition of more litharge the mass was brought to a round bead; though with considerable difficulty. This bead bore melting on the charcoal, without any reduction of the lead, but could not be obtained transparent.

(E) The ease with which this substance had melted with vegetable ashes, led to the trial of it with pure calcareous earth. A fragment of Tabasheer, fixed to the end of a bit of glass, was rubbed over with some powdered whiting. As soon as exposed to the flame of the blow-pipe, it melted with considerable effervescence; but could not, even on the charcoal, and with the addition of more whiting, be brought to a transparent state, or reduced into a round bead.

Equal weights of Tabasheer and pure calcareous spar, both reduced to fine powder, were irregularly mixed, and exposed in the platina crucible to a strong fire in a forge for 20'; but did not even concrete together.

(F) When magnesia was used, no fusion took place at the blow-pipe.

(G) Equal parts of Tabasheer, whiting, and earth of alum precipitated by mild volatile alkali, were mixed in a state of powder, and submitted in the platina crucible to a strong fire for 20', but were afterwards found unmelted.

Examination of the other specimens.

No. I.

This parcel contained particles of three kinds; some white, of a smooth texture, much resembling the foregoing sort; others of the same appearance, but yellowish; and others greatly similar to bits of dried mould.

The white and yellowish pieces were so soft as to be very

easily rubbed to powder between the fingers. They had a disagreeable taste, something like that of rhubarb. Put into water, the white bits scarcely grew at all transparent; but the yellow ones became so to a considerable degree.

The brown earth-like pieces were harder than the above, had little taste, floated upon water, and remained opaque.

Exposed to the blow-pipe, they all charred and grew black; the last variety even burned with a flame. When the vegetable matter was consumed, the pieces remained white, and then had exactly the appearance, and possessed all the properties, of the foregoing Tabasheer from Hydrabad, and like it melted with soda into a transparent glass.

No. II.

Also consisted of bits of three sorts.

(a) Some white, nearly opaque.

(b) A few small very transparent particles, shewing, in an eminent degree, the blue and yellow colour, by the different direction of light.

(c) Coarse, brownish pieces of a grained texture.

These all had exactly the same taste, hardness, &c., and shewed the same effects at the blow-pipe, as No. I.

27 gr. of this Tabasheer thrown into a red-hot crucible, burned with a yellowish white flame, lost 2.9 gr. in weight, and became so similar to the Hydrabad kind as not to be distinguished from it.

Some of this Tabasheer put into a crucible, not made very hot emitted a smell something like tobacco ashes, but not the kind of perfume discovered in that from Hydrabad, § IV. (E).

No. IV.

All the pieces of this parcel were of one appearance, and a good deal resembled, in their texture, the third variety of No. II. Their colour was white; their hardness such as very difficultly to be broken by pressure between the fingers.

In the mouth they immediately fell to a pulpy powder, and had no taste.

A bit exposed on the charcoal to the blow-pipe became black, melted like some vegetable matters, caught flame, and burnt to a botryoid inflated coal, which soon entirely consumed away, and vanished.

A piece put into water fell to a powder.   The mixture being boiled, this powder dissolved, and turned the whole to a jelly.

These properties are exactly those of common starch.

No. V.

Agreed entirely with No. IV. in appearance, properties, and nature.

No. VI.

The pieces of this parcel were white, quite opaque, and considerably hard.   Their taste and effects at the blow-pipe, were perfectly similar to those of the Hydrabad kind.

No. VII.

Much resembled No. VI. only was rather softer, and seemed to blacken a little when first heated.   With fluxes at the blow-pipe it shewed the same effects as all the above.

Conclusion.

1. It appears from these experiments, that all the parcels, except No. IV. and V. consisted of genuine Tabasheer; but that those kinds, immediately taken from the plant, contained a certain portion of a vegetable matter, which was wanting in the specimens procured from the shops, and which had probably been deprived of this admixture by calcination, of which operation a partial blackness, observable on some of the pieces of No. III. and VI. are doubtless the traces.   This accounts also for the superior hardness and diminished tastes of these sorts.

2. The nature of this substance is very different from what might have been expected in the product of a vegetable. Its indestructibility by fire; its total resistance to acids; its uniting by fusion with alkalies in certain proportions into a white opaque mass, in others into a transparent permanent glass; and its being again separable from these compounds, entirely unchanged by acids, &c., seem to afford the strongest reasons to consider it as perfectly identical with common *siliceous earth.*

Yet from pure quartz it may be thought to differ in some material particulars; such as in its fusing with calcareous earth, in some of its effects with liquid alkalies, in its taste, and its specific gravity.

But its taste may arise merely from its divided state, for chalk and powdery magnesia both have tastes, and tastes which are very similar to that of pure Tabasheer; but when these earths are taken in the denser state of crystals, they are found to be quite insipid; so Tabasheer, when made more solid by exposure to a pretty strong heat, is no longer perceived, when chewed, to act upon the palate, § IV. (A). ·

And, on accurate comparison, its effects with liquid alkalies have not appeared peculiar; for though it was found on trial, that the powder of common flints, when boiled in some of the same liquid caustic alkali employed at § IX. (A) was scarcely at all acted upon; and that the very little which was dissolved, was soon precipitated again, in the form of minute *flocculi*, on exposing the solution to the air, and was immediately thrown down on the admixture of an acid; yet the precipitate obtained from *liquor silicum* by marine acid was discovered, even when dry to dissolve readily in this alkali, but while still moist to do so very copiously, even without the assistance of heat; and some of this solution, thus saturated with siliceous matter by ebullition, being exposed to the air in a shallow glass, became a jelly by the next day, and the day after dried, and cracked, &c., exactly like the mixtures § IX. (D and E). And another portion of this solution mixed with marine acid afforded no precipitate, and remained perfectly unaffected for two days; but on the third it was converted into a firm jelly like that § IX. (F).

As gypsum is found to melt *per se* at the blow-pipe, though refractory to the strongest heat that can be made in a furnace, it was thought that possibly siliceous and calcareous earths might flux together by this means, though they resist the utmost power of common fires; but experiment showed that in this respect quartz did not agree with Tabasheer. But this difference seems much too likely to depend on the admixture of a little foreign matter in the latter body, to admit of its being made the grounds for considering it as a new substance, in opposition to so many more material points in which it agrees with silex.

Nor can much weight be laid on the inferior specific gravity of a body so very porous. The infusibility of the mixture § XIII. (G) depended also, probably, either on an inaccuracy in the proportions of the earths to each other, or on a deficiency of heat.

. 3. Of the three bamboos which were not split before the Royal Society, I have opened two. The Tabasheer found in them agreed entirely in its properties with that of No. I. and II.

It was observed that all the Tabasheer in the same joint was exactly of the same appearance. In one joint it was all similar to the yellowish sort No. I. In another joint of the same bamboo, it resembled the variety (c) of No. II. Probably, therefore, the parcels from Dr. RUSSELL, containing each several varieties of this substance, arose from the produce of many joints having been mixed together.

4. The ashes, obtained by burning the bamboo, boiled in marine acid, left a very large quantity of a whitish insoluble powder, which, fused at the blow-pipe with soda, effervesced and formed a transparent glass. Only the middle part of the joints was burned, the knots were sawed off, lest being porous, Tabasheer might be mechanically lodged in them. However, the great quantity of this remaining

substance shews it to be an essential, constituent part of the wood.

The ashes of common charcoal, digested in marine acid, left in the same manner an insoluble residuum which fused with soda with effervescence, and formed glass; but the proportion of this matter to the ashes was greatly less than in the foregoing case.

5. Since the above experiments were made, a singular circumstance has presented itself. A green bamboo, cut in the hot-house of Dr. Pitcairn, at Islington, was judged to contain Tabasheer in one of its joints, from a rattling noise discoverable on shaking it; but being split by Sir Joseph Banks, it was found to contain, not ordinary Tabasheer, but a solid pebble, about the size of half a pea.

Externally this pebble was of an irregular rounded form, of a dark-brown or black colour. Internally it was reddish brown, of a close dull texture, much like some martial siliceous stones. In one corner there were shining particles, which appeared to be crystals, but too minute to be distinguished even with the microscope.

This substance was so hard as to cut glass!

A fragment of it exposed to the blow-pipe on the charcoal did not grow white, contract in size, melt, or undergo any change. Put into borax it did not dissolve, but lost its colour, and tinged the flux green. With soda it effervesced, and formed a round bead of opaque black glass.

These two beads, digested in some perfectly pure and white marine acid, only partially dissolved, and tinged this menstruum of a greenish yellow colour; and from this solution Prussite of tartar, so pure as not, under many hours, to produce a blue colour with the above pure marine acid, instantly threw down a very copious Prussian blue.

P. S.—In ascertaining the specific gravity of the Hydrabad Tabasheer, § I. (G), great care was taken in both the experiments that every bit was thoroughly penetrated with the water, and transparent to its very centre, before its weight in the water was determined.

2

A CHEMICAL ANALYSIS OF SOME CALAMINES.

From the Philosophical Transactions of the Royal Society of London,
Vol. XCIII, page 12.—Read November, 18, 1802.

Notwithstanding the experiments of BERGMAN and others, on those ores of zinc which are called calamine, much uncertainty still subsisted on the subject of them. Their constitution was far from decided, nor was it even determined whether all calamines were of the same species, or whether there were several kinds of them.

The Abbé HAÜY, so justly celebrated for his great knowledge in crystallography and mineralogy, has adhered, in his late work,* to the opinions he had before advanced,† that calamines were all of one species, and contained no carbonic acid, being a simple calx of zinc, attributing the effervescence which he found some of them to produce with acids, to an accidental admixture of carbonate of lime.

The following experiments were made to obtain a more certain knowledge of these ores; and their results will show the necessity there was for their farther investigation, and how wide from the truth have been the opinions adopted concerning them.

Calamine from Bleyberg.

a. The specimen which furnished the subject of this article, was said by the German of whom it was purchased, to have come from the mines of Bleyberg in Carinthia.

It was in the form of a sheet stalactite, spread over small fragments of limestone. It was not however at all crystalline, but of the dull earthy appearance of chalk, though, on comparison, of a finer grain and closer texture.

It was quite white, perfectly opaque, and adhered to the

---

* *Traité de Mineralogie*, Tome IV.   † *Journal des Mines.*

tongue; 68.0 grs. of it, in small bits, immersed in distilled water, absorbed 19.8 grs. of it, $= 0.29$.

It admitted of being scraped by the nail though with some difficulty: scraped with a knife, it afforded no light.

68.1 grs. of it, broken into small pieces, expelled 19.0 grs. of distilled water from a stopple bottle. Hence its density $= 3.584$. In another trial, 18.96 grs. at a heat of 65° FAHRENHEIT, displaced 5.27 grs. of distilled water; hence the density $= 3.598$. The bits, in both cases, were entirely penetrated with water.

b. Subjected to the action of the blowpipe on the coal, it became yellow the moment it was heated, but recovered its pristine whiteness on being let cool. This quality, of temporarily changing their colour by heat, is common to most, if not all, metallic oxides; the white growing yellow, the yellow red, the red black.

Urged with the blue flame, it became extremely friable; spread yellow flowers on the coal; and, on continuing the fire no very long time, entirely exhaled. If the flame was directed against the flowers, which had settled on the coal, they shone with a vivid light. A bit fixed to the end of a slip of glass, wasted nearly as quickly as on the coal.

It dissolved in borax and microcosmic salt, with a slight effervescence, and yielded clear colourless glasses; but which became opaque on cooling, if over saturated. Carbonate of soda had not any action on it.

c. 68.0 grs. of this calamine dissolved in dilute vitriolic acid with a brisk effervescence, and emitted 9.2 grs. of carbonic acid. The solution was white and turbid, and on standing deposited a white powder, which, collected on a small filter of gauze paper, and well edulcorated and let dry, weighed only 0.86 gr. This sediment, tried at the blowpipe, melted first into an opaque white matter, and then partially reduced into lead. It was therefore, probably, a mixture of vitriol of lead and vitriol of lime.

The filtered solution, gently exhaled to dryness, and kept over a spirit-lamp till the water of crystallization of the

salt and all superfluous vitriolic acid were driven off, afforded 96.7 grs. of perfectly dry, or *arid*,* white salt. On
re-solution in water, and crystallization, this saline matter
proved to be wholly vitriol of zinc, excepting an inappre-
tiable quantity of vitriol of lime in capillary crystals, due,
without doubt, to a slight and accidental admixture of some
portion of the calcareous fragments on which this calamine
had been deposited. Pure martial prussiate of tartar,
threw down a white precipitate from the solution of this
salt.

In another experiment, 20.0 grs. of this calamine afforded
28.7 grs. of arid vitriol of zinc.

d. 10 grs. of this calamine were dissolved in pure marine
acid, with heat. On cooling, small capillary crystals of
muriate of lead formed in the solution. This solution was
precipitated by carbonate of soda, and the filtered liquor let
exhale slowly in the air; but it furnished only crystals of
muriate of soda.

e. 10 grs. dissolved in acetous acid without leaving any
residuum. By gentle evaporation, 20.3 grs. = 2.03, of ace-
tite of zinc, in the usual hexagonal plates, were obtained.
These crystals were permanent in the air, and no other
kind of salt could be perceived amongst them.

Neither solution of vitriolated tartar, nor vitriolic acid,
occasioned the slightest turbidness in the solution of these
crystals, either immediately or on standing; a proof that
the quantity of lime and lead in this solution, if any, was
excessively minute.

f. A bit of this calamine, weighing 20.6 grs. being made
red hot in a covered tobacco-pipe, became very brittle, di-
viding on the slightest touch into prisms, like those of
starch, and lost 5.9 grs. of its weight = 0.286. After this,
it dissolved slowly and difficultly in vitriolic acid, without
any effervescence.

---

* *Dry*, as opposed to wet or damp, which are only degrees of each other,
merely implies free from mechanically admixed water. *Arid*, may be ap-
propriated to express the state of being devoid of combined water.

According to these experiments, this calamine consists of

Calx of zinc	-	-	-	0.714
Carbonic acid	-	-	-	0.135
Water	-	-	-	0.151

$$\overline{1.000.}$$

The carbonates of lime and lead in it are mere accidental admixtures, and in too small quantity to deserve notice.

Calamine from Somersetshire.

a. This calamine came from Mendip Hills in Somersetshire.

It had a mammillated form; was of a dense crystalline texture; semitransparent at its edges, and in its small fragments; and upon the whole very similar, in its general appearance, to calcedony.

It was tinged, exteriorly, brown; but its interior colour was a greenish yellow.

It had considerable hardness; it admitted however of , being scraped by a knife to a white powder.

56.8 grs. of it displaced 13.1 grs. of water, at a temperature of 65° FAHRENHEIT. Hence its density = 4.336.

b. Exposed to the blowpipe, it became opaque, more yellow, and friable; spread flowers on the coal, and consequently volatilized, but not with the rapidity of the foregoing kind from Bleyberg.

It dissolved in borax and microcosmic salt, with effervescence, yielding colourless glasses. Carbonate of soda had no action on it.

c. It dissolved in vitriolic acid with a brisk effervescence; and 67.9 grs. of it emitted 24.5 grs. = 0.360, of carbonic acid. This solution was colourless; and no residuum was left. By evaporation, it afforded only vitriol of zinc, in pure limpid crystals.

d. 23.0 grs. in small bits, made red hot in a covered· tobacco-pipe, lost 8.1 grs. = 0.352. It then dissolved slowly

and difficultly in vitriolic acid, without any emission of carbonic acid; and, on gently exhaling the solution, and heating the salt obtained, till the expulsion of all superabundant
vitriolic acid and all water, 29.8 grs. of arid vitriol of zinc
were obtained. This dry salt was wholly soluble again in
water; and solution of pure martial prussiate of soda occasioned a white precipitate in it.

This calamine hence consists of

Carbonic acid	-	-	-	0.352
Calx of zinc	-	-	-	0.648
				1.000.

Calamine from Derbyshire.

a. This calamine consisted of a number of small crystals,
about the size of tobacco-seeds, of a pale yellow colour,
which appeared, from the shape of the mass of them, to
have been deposited on the surface of crystals of carbonate
of lime, of the form of Fig. 28, Plate IV. of the *Cristallographie* of ROMÉ DE L'ISLE.

The smallness of these calamine crystals, and a want of
sharpness, rendered it impossible to determine their form
with certainty; they were evidently, however, rhomboids,
whose faces were very nearly, if not quite, rectangular, and
which were incomplete along their six intermediate edges,
apparently like Fig. 78, Plate IV. of ROMÉ DE L'ISLE.

22.1 grs. of these crystals, at a heat of 57° FAHRENHEIT,
displaced 5.1 grs. of water, which gives their density =
4.333.

Heat did not excite any electricity in these crystals.

b. Before the blowpipe, they grew more yellow and
opaque, and spread flowers on the coal. They dissolved
wholly in borax and microcosmic salt, with effervescence.

c. 22.0 grs. during their solution in vitriolic acid, effervesced, and lost 7.8 grs. of carbonic acid = 0.354. This
solution was colourless, and afforded 26.8 grs. of arid vitriol
of zinc, which, redissolved in water, shot wholly into clear
colourless prisms of this salt.

d. 9.2 grs. of these crystals, ignited in a covered tobacco-pipe, lost 3.2 grs. $=$ 0.3478; hence, these crystals consist of

Carbonic acid	- - -	0.348
Calx of zinc	- - -	0.652
		1.000.

Electrical Calamine.

The Abbé HAÜY has considered this kind as differing from the other calamines only in the circumstance of being in distinct crystals; but it has already appeared, in the instance of the Derbyshire calamine, that all crystals of calamine are not electric by heat, and hence, that it is not merely to being in this state that this species owes the above quality. And the following experiments, on some crystals of electric calamine from Regbania in Hungary, can leave no doubt of its being a combination of calx of zinc with quartz; since the quantity of quartz obtained, and the perfect regularity and transparency of these crystals, make it impossible to suppose it a foreign admixture in them.

a. 23.45 grs. of these Regbania crystals, displaced 6.8 grs. of distilled water, from a stopple-bottle, at the temperature of 64° FAHRENHEIT; their specific gravity is therefore $=$ 3.434.

The form of these crystals is represented in the annexed Figure.

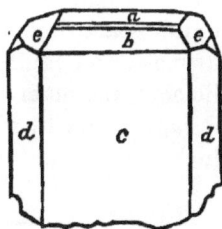

$$a\,c = 90°.$$
$$a\,e = 150°.$$
$$b\,c = 115°.$$
$$c\,d = 130°.$$

They were not scratched by a pin; a knife marked them.

b. One of these crystals, exposed to the flame of the blow-pipe, decrepitated and became opaque, and shone with a green light, but seemed totally infusible.

Borax and microcosmic salt dissolved these crystals, without any effervescence, producing clear colourless glasses. Carbonate of soda had little if any action on them.

c. According to Mr. PELLETIER's experiments* on the calamine of Fribourg in Brisgaw, which is undoubtedly of this species, its composition is,

Quartz - - -	0.50
Calx of zinc - - -	0.38
Water - - -	0.12
	1.00.

The experiments on the Regbania crystals have had different results; but, though made on much smaller quantities, they will perhaps not be found, on repetition, less in conformity with nature.

23.45 grs. heated red hot in a covered crucible, decrepitated a little, and became opaque, and lost 1.05 gr. but did not fall to powder or grow friable. It was found that this matter was not in the least deprived of its electrical quality by being ignited; and hence, while hot, the fragments of these decrepitated crystals clung together, and to the crucible.

d. 22.2 grs. of these decrepitated crystals, = 23.24 grs. of the original crystals, in a state of impalpable powder, being digested over a spirit-lamp with diluted vitriolic acid, showed no effervescence; and after some time, the mixture became a jelly. Exhaled to dryness, and ignited slightly, to expel the superfluous vitriolic acid, the mass weighed 37.5 grs.

On extraction of the saline part by distilled water, a fine powder remained, which, after ignition, weighed 5.8 grs. and was quartz.

---

* *Journal de Physique,* Tome **XX.** p 424.

The saline solution afforded on crystallization, only vitriol of zinc.

These crystals therefore consist of

Quartz	-	-	-	0.250
Calx of zinc	-	-	-	0.683
Water	-	-	-	0.044
				0.977
Loss	-	-	-	0.023
				1.000.

The water is most probably not an essential element of this calamine, or in it in the state of, what is improperly called, water of crystallization, but rather exists in the crystals in fluid drops interposed between their plates, as it often is in crystals of nitre, of quartz, &c. Its small quantity, and the crystals not falling to powder on its expulsion, but retaining almost perfectly their original solidity, and spathose appearance in the places of fracture, and, above all, preserving their electrical quality wholly unimpaired, which would hardly be the case after the loss of a real element of their constitution, seem to warrant this opinion.

If the water is only accidental in this calamine, its composition, from the above experiments, will be

Quartz	-	-	-	0.261
Calx of zinc	-	-	-	0.739
				1.000.

I have found this species of calamine amongst the productions of Derbyshire, in small brown crystals, deposited, together with the foregoing small crystals of carbonate of zinc, on crystals of carbonate of lime. Their form seems, as far as their minuteness and compression together would allow of judging, nearly or quite the same as that of those from Regbania; and the least atom of them immediately evinces its nature, on being heated, by the strong electricity it acquires. On their solution in acids, they leave quartz.

OBSERVATIONS.

Chemistry is yet so new a science, what we know of it bears so small a proportion to what we are ignorant of, our knowledge in every department of it is so incomplete, so broken, consisting so entirely of isolated points thinly scattered like lurid specks on a vast field of darkness, that no researches can be undertaken without producing some facts, leading to some consequences, which extend beyond the boundaries of their immediate object.

1. The foregoing experiments throw light on the proportions in which its elements exist in vitriol of zinc. 23.0 grs. of the Mendip Hill calamine, produced 29.8 grs. of arid vitriol of zinc. These 23.0 grs. of calamine contained 14.9 grs. of calx of zinc; hence, this metallic salt, in an arid state, consists of *exactly equal* parts of calx of zinc and vitriolic acid.

This inference is corroborated by the results of the other experiments: 68.0 grs. of the Bleyberg calamine, containing 48.6 grs. of calx of zinc, yielded 96.7 grs. of arid vitriol of zinc; and, in another trial, 20.0 grs. of this ore, containing 14.2 grs. of calx of zinc, produced 28.7 grs. of arid vitriol of zinc. The mean of these two cases, is 62.7 grs. of arid vitriol of zinc, from 31.4 grs. of calx of zinc.

In the experiment with the crystals of carbonate of zinc from Derbyshire, 14.35 grs. of calx of zinc furnished indeed only 26.8 grs. of arid vitriol of zinc; a deficiency of about $\frac{8}{100}$, occasioned probably by some small inaccuracy of manipulation.

2. When the simplicity found in all those parts of nature which are sufficiently known to discover it is considered, it appears improbable that the proximate constituent parts of bodies should be united in them, in the very remote relations to each other in which analyses generally indicate them; and, an attention to the subject has led me to the opinion that such is in fact not the case, but that, on the contrary, they are universally, as appears here with respect

to arid vitriol of zinc, fractions of the compound of very low denominators. Possibly in few cases exceeding five.

The success which has appeared to attend some attempts to apply this theory, and amongst others, to the compositions of some of the substances above analysed, and especially to the calamine from Bleyberg, induces me to venture to dwell here a little on this subject, and state the composition of this calamine which results from the system, as, besides contributing perhaps to throw some light on the true nature of this ore, it may be the means likewise of presenting the theory under circumstances of agreement with experiment, which from the surprising degree of nearness, and the trying complexity of the case, may seem to entitle it to some attention.

From this calamine, containing, according to the results of the experiments on the Mendip Hill kind, too small a quantity of carbonic acid to saturate the whole of the calx of zinc in it, and from its containing much too large a portion of water to be in it in the state of mere moisture or dampness, it seems to consist of two matters ; carbonate of zinc, and a peculiar compound of zinc and water, which may be named *hydrate of zinc.*

By the results of the analysis of the Mendip Hill calamine, corrected by the theory, carbonate of zinc appears to consist of

Carbonic acid	-	-	-	$\frac{1}{3}$
Calx of zinc	-	-	-	$\frac{2}{3}$

Deducting from the calx of zinc in the Bleyberg calamine, that portion which corresponds, on these principles, to its yield of carbonic acid, the remaining quantity of calx of zinc and water are in such proportions as to lead, from the theory, to consider hydrate of zinc as composed of

Calx of zinc	-	-	-	$\frac{3}{4}$
Water, or rather ice	-	-		$\frac{1}{4}$

And, from these results, corrected by the theory, I consider Bleyberg calamine as consisting of

Carbonate of zinc	-	-	-	$\frac{2}{3}$
Hydrate of zinc	-	-	-	$\frac{1}{3}$

The test of this hypothesis, is in the quantities of the remote elements which analysis would obtain from a calamine thus composed.

The following table will show how very insignificantly the calamine compounded by the theory, would differ in this respect from the calamine of nature.

1000 parts of the compound salt of carbonate and hydrate of zinc consist of

$$
\begin{array}{ll}
\text{Carbonate of zinc } 400 = \left\{
\begin{array}{l}
\text{Carbonic acid } = \dfrac{400}{3} = \qquad \cdots \qquad - 133\tfrac{1}{3} \\[2ex]
\text{Calx of zinc } = \dfrac{400 \times 2}{3} = 266\tfrac{2}{3}
\end{array}
\right. \\[5ex]
\text{Hydrate of zinc } = 600 \left\{
\begin{array}{l}
\text{Calx of zinc } = \dfrac{600 \times 3}{4} = 450 \\[2ex]
\text{Ice } \cdots = \dfrac{600}{4} = \qquad \cdots \qquad 150
\end{array}
\right.
\end{array}
$$

$$= \cdots - 716\tfrac{2}{3}$$

$$\overline{1000.}$$

Great as is the agreement between the quantities of the last column and those obtained by the analysis of the Bleyberg calamine, it would be yet more perfect, probably, had there been, in this instance, no sources of fallacy but those attached to chemical operations, such as errors of weighing, waste, &c., but the differences which exist are owing, in some measure at least, to the admixture of carbonate of lime and carbonate of lead, in the calamine analysed, and also to some portion of water, which is undoubtedly contained, in the state of moisture, in so porous and bibulous a body.

It has also appeared, in the experiments on the Mendip Hill calamine, that acids indicate a greater quantity of carbonic acid than fire does, by $\frac{22}{1000}$. If we make this deduction for dissolved water, it reduces the quantity of carbonic acid in the Bleyberg calamine, to 0.1321.

If we assume this quantity of carbonic acid as the datum to calculate, on this system, the composition of the calamine from Bleyberg, we shall obtain the following results:

Compound salt, of carbonate of zinc and hydrate
of zinc   -   -   -   -   -   990.3
Water in the state of moisture   -   -   2.5
Carbonate of lime and carbonate of lead   -   7.2

————

1000.0

It may be thought some corroboration of the system here offered, that, if we admit the proportions which it indicates, the remote elements of this ore, while they are regular parts of their immediate products, by whose subsequent union this ore is engendered, are also regular fractions of the ore itself: thus,

The carbonic acid   -   - $= \frac{8}{80}$

The water   -   -   - $= \frac{9}{80}$

The calx of zinc   -   - $= \frac{43}{80}$

Hereby displaying that sort of regularity, in every point of view of the object, which so wonderfully characterises the works of nature, when beheld in their true light.

If this calamine does consist of carbonate of zinc and hydrate of zinc, in the regular proportions above supposed, little doubt can exist of its being a true chemical combination of these two matters, and not merely a mechanical mixture of them in a pulverulent state; and, if so, we may indulge the hope of some day meeting with this ore in regular crystals.

If the theory here advanced has any foundation in truth the discovery will introduce a degree of rigorous accuracy and certainty into chemistry, of which this science was thought to be ever incapable, by enabling the chemist, like the geometrician, to rectify by calculation the unavoidable errors of his manual operations, and by authorising him to eliminate from the essential elements of a compound, those products of its analysis whose quantity cannot be reduced to any admissible proportion.

A certain knowledge of the exact proportions of the constituent principles of bodies, may likewise open to our view harmonious analogies between the constitutions of related objects, general laws, &c., which at present totally escape us. In short, if it is founded in truth, its enabling the application of mathematics to chemistry, cannot but be productive of material results.*

3. By the application of the foregoing theory to the experiments on the electrical calamine, its elements will appear to be,

Quartz   -   -   -   - $\frac{1}{4}$
Calx of zinc   -   -   - $\frac{3}{4}$

A small quantity of the calamine having escaped the action of the vitriolic acid, and remained undecomposed, will account for the slight excess in the weight of the quartz.

4. The exhalation of these calamines at the blowpipe, and the flowers which they diffuse round them on the coal, are probably not to be attributed to a direct volatilization of them. It is more probable that they are the consequences of the disoxidation of the zinc calx, by the coal and the inflammable matter of the flame, its sublimation in a metallic state, and instantaneous recalcination. And this alternate reduction and combustion, may explain the peculiar phosphoric appearance exhibited by calces of zinc at the blowpipe.

The apparent sublimation of the common flowers of zinc at the instant of their production, though totally unsublimable afterwards, is certainly likewise but a deceptious appearance. The reguline zinc, vaporized by the heat, rises from the crucible as a metallic gas, and is, while in this state, converted to a calx. The flame which attends the process is a proof of it; for flame is a mass of vapour, ignited by the production of fire within itself. The fibrous form of the flowers of zinc, is owing to a crystallization of the calx while in *mechanical suspension* in the air, like that which takes place with camphor, when, after having been sometime inflamed, it is blown out.

A moment's reflection must evince, how injudicious is the

---

* It may be proper to say, that the experiments have been stated *precisely* as they turned out, and have not been in the *least degree* bent to the system.

common opinion, of crystallization requiring a state of solution in the matter; since it must be evident, that while solution subsists, as long as a quantity of fluid admitting of it is present, no crystallization can take place. The only requisite for this operation, is a freedom of motion in the masses which tend to unite, which allows them to yield to the impulse which propels them together, and to obey that sort of polarity which occasions them to present to each other the parts adapted to mutual union. No state so completely affords these conditions as that of mechanical suspension in a fluid whose density is so great, relatively to their size, as to oppose such resistance to their descent in it as to occasion their mutual attraction to become a power superior to their force of gravitation. It is in these circumstances that the atoms of matters find themselves, when, on the separation from them of the portion of fluid by which they were dissolved, they are abandoned in a disengaged state in the bosom of a solution; and hence it is in saturated solutions sustaining evaporation, or equivalent cooling, and free from any perturbing motion, that regular crystallization is usually effected.

But those who are familiar with chemical operations, know the sort of agglutination which happens between the particles of subsided very fine precipitates; occasioning them, on a second diffusion through the fluid, to settle again much more quickly than before, and which is certainly a crystallization, but under circumstances very unfavourable to its perfect performance.

5. No calamine has yet occurred to me which was a real, uncombined, calx of zinc. If such, as a native product, should ever be met with in any of the still unexplored parts of the earth, or exist amongst the unscrutinized possessions of any cabinet, it will easily be known, by producing a quantity of arid vitriol of zinc exactly double its own weight; while the hydrate of zinc, should it be found single, or uncombined with the carbonate, will yield, it is evident, 1.5 its weight of this arid salt.

ACCOUNT OF A DISCOVERY OF NATIVE MINIUM.

From the Philosophical Transactions of the Royal Society of London,
Vol. XCVI, Part I, 1806, p. 267.—Read April 24, 1806.

IN A LETTER TO THE RIGHT HON. SIR JOSEPH BANKS, K. B. P. R. S.

My Dear Sir : I beg leave to acquaint you with a discovery which I have lately made, as it adds a new, and perhaps it may be thought an interesting, species to the ores of lead. I have found *minium* native in the earth.

It is disseminated in small quantity, in the substance of a compact carbonate of zinc.

Its appearance in general is that of a matter in a pulverulent state, but in places it shows to a lens a flaky and crystalline texture.

Its colour is like that of factitious minium, a vivid red with a cast of yellow.

Gently heated at the blowpipe it assumes a darker colour, but on cooling it returns to its original red. At a stronger heat it melts to litharge. On the charcoal it reduces to lead.

In dilute white acid of nitre, it becomes of a coffee colour. On the addition of a little sugar, this brown calx dissolves, and produces a colourless solution.

By putting it into marine acid with a little leaf gold, the gold is soon intirely dissolved.

When it is inclosed in a small bottle with marine acid, and a little bit of paper tinged by turnsol is fixed to the cork, the paper in a short time entirely loses its blue colour, and becomes white. A strip of common blue paper, whose colouring matter is indigo, placed in the same situation undergoes the same change.

The very small quantity which I possess of this ore, and the manner in which it is scattered amongst another substance, and blended with it, have not allowed of more qualities being determined, but I apprehend these to be sufficient to establish its nature.

This native minium seems to be produced by the decay of a galena, which I suspect to be itself a secondary production from the metallization of white carbonate of lead by hepatic gas. This is particularly evident in a specimen of this ore which I mean to send to Mr. GREVILLE, as soon as I can find an opportunity. In one part of it there is a cluster of large crystals. Having broken one of these, it proved to be converted into minium to a considerable thickness, while its centre is still galena.

I am, &c.,

JAMES SMITHSON.

CASSELL IN HESSE, *March 2d*, 1806.

From the Philosophical Magazine, Vol. XXXVIII, 1811, p. 84.

After I had communicated to the president the account of the discovery of native minium, printed in the Philosophical Transactions for 1806, I learned that this ore came from the lead mines of Breylau in Westphalia.

ON QUADRUPLE AND BINARY COMPOUNDS, PARTICULARLY SULPHURETS.

From the Philosophical Magazine, London, Vol. XXIX, 1807, p. 275. Read December 24, 1807.

A paper, by Mr. Smithson, on quadruple and binary compounds, particularly the sulphurets, was read. The author seemed to doubt the propriety of the distinction, or rather the existence, of quadruple compounds, believed that only two substances could enter as elements in the composition of one body, and contended that in cases of quadruple compounds, a new and very different substance was formed, which had very little relation to the radical or elementary principles of which it was believed to be composed. This opinion he supported by reference to the sulphurets of lead (galena) and of antimony, and to the facts developed by crystallography. In the latter science he took occasion to correct and confirm some remarks of his in the Transactions for 1804, on different crystals, which he acknowledged have not hitherto been found in nature.

ON THE COMPOSITION OF THE COMPOUND SULPHURET FROM HUEL BOYS, AND AN ACCOUNT OF ITS CRYSTALS.

From the Philosophical Transactions of the Royal Society of London, Vol. XCVIII, Part I, 1808, p. 55.—Read January 28, 1808.

It is but very lately that I have seen the Philosophical Transactions for 1804, and become acquainted with the two papers on the compound sulphuret of lead, antimony, and copper contained in the first part of it, which circumstance has prevented my offering sooner a few observations on Mr. HATCHETT's experiments, which I deem essential towards this substance being rightly considered, and indeed the principles of which extend to other chemical compounds; and also giving an account of the form of this compound sulphuret, as that which has been laid before the Society is very materially inaccurate and imperfect.

We have no real knowledge of the nature of a compound substance till we are acquainted with its proximate elements, or those matters by whose direct or immediate union it is produced; for these only are its true elements. Thus, though we know that vegetable acids consist of oxygene, hydrogene, and carbon, we are not really acquainted with their composition, because these are not their proximate, that is, are not their elements, but are the elements of their elements, or the elements of these. It is evident what would be our acquaintance with sulphate of iron; for example, did we only know that a crystal of it consisted of iron, sulphur, oxygene, and hydrogene; or of carbonate of lime, if only that it was a compound of lime, carbon or diamond, and oxygene. In fact, totally dissimilar substances may have the same ultimate elements, and even probably in precisely the same proportions; nitrate of ammonia, and hydrate of ammonia, or crystals of caustic volatile alkali,* both ultimately consist of oxygene, hydrogene, and azote.

It is not probable that the present ore is a direct quadruple combination of the three metals and sulphur, that these, in their simple states, are its immediate component parts; it is much more credible that it is a combination of the three sulphurets of these metals.

On this presumption I have made experiments to determine the respective proportions of these sulphurets in it.

I have found 10 grains of galena, or sulphuret of lead, to produce 12.5 grains of sulphate of lead. Hence the 60.1

---

* FOURCROY, Syst. des Con. Chem. t. I. p. 88.

grains of sulphate lead, which Mr. HATCHETT obtained, correspond to 48.08 grains of sulphuret of lead.

I have found 10 grains of sulphuret of antimony to afford 11.0 grains of precipitate from muriatic acid by water. Hence 31.5 grains of this precipitate are equal to 28.64 grains of sulphuret of antimony.

The want of sulphuret of copper has prevented my determining the relation between it and black oxide of copper, but this omission is, it is evident, immaterial, as the quantity of this sulphuret in the ore must be the complement of the sum of the two others.

But as the iron is a foreign adventitious substance in this ore, it follows that the foregoing quantities are the products of only 96.65 grains of it. 100 parts of the ore are therefore composed of

Sulphuret of lead	- -	49.7
Sulphuret of antimony	- -	29.6
Sulphuret of copper	- -	20.7
		100.0

It is impossible not to be struck with the trifling alteration which these quantities require to reduce them to very simple proportions, or to think it a very great violation of probability to suppose that experiments, effected with no errors, would have given them thus:

Sulphuret of lead	- - -	50.
Sulphuret of antimony	- -	30.
Sulphuret of copper	- - -	20.

However, I doubt the existence of triple, quadruple, &c. compounds; I believe, that *all combination is binary;* that no substance whatever has more than two proximate or true elements; and hence I should be inclined to consider the present compound as a combination of galena and fahlertz; and if so, it will be accurately represented, as far as chemical analysis has yet been able to go, by the following figure:

$$\text{Compound sulphuret of lead, antimony, and copper} = \begin{cases} \tfrac{1}{2}\text{ galena} = \begin{cases} \tfrac{1}{6}\text{ sulphur} \\ \tfrac{5}{6}\text{ lead} \end{cases} \\ \tfrac{1}{2}\text{ fahlertz} = \begin{cases} \tfrac{4}{5}\text{ sulphuret of antimony} = \begin{cases} \tfrac{1}{2}\text{ sulphur.} \\ \tfrac{1}{2}\text{ antimony.} \end{cases} \\ \tfrac{1}{5}\text{ sulphuret of copper} = \begin{cases} \tfrac{1}{4}\text{ sulphur.} \\ \tfrac{3}{4}\text{ copper.} \end{cases} \end{cases} \end{cases}$$

Its ultimate elements are therefore,

Sulphur	-	-	$20 \ldots = \tfrac{12}{60}$
Lead	-	-	$41\tfrac{2}{3} \ldots = \tfrac{25}{60}$
Antimony	-	-	$25 \ldots = \tfrac{15}{60}$
Copper	-	-	$13\tfrac{1}{3} \ldots = \tfrac{8}{60}$

and it is not a little remarkable, that here, as was the case with the calamine,* they are sexagesimal fractions of it.

When in a former paper I offered a system on the proportions of the elements of compounds, I supported it by the results of my own experiments, which might be supposed influenced, even unconsciously to myself, by a favourite hypothesis, and I made the application of it principally to a substance whose nature was not very clear. But the present case is not liable to these objections : here no fondness to the theory can be suspected of having led astray, nor did even the experiments as they came from their author's hands, bear an appearance in the least favourable to it, and yet when properly considered, they are found to accord no less remarkably with its principles.

It is evident that there must be a precise quantity in which the elements of compounds are united together in them, otherwise a matter, which was not a simple one, would be liable, in its several masses, to vary from itself, according as one or other of its ingredients chanced to predominate; but chemical experiments are unavoidably attended with too many sources of fallacy for this precise quantity to be discovered by them; it is therefore to theory

---

* Phil. Trans. 1803, p. 12.

that we must owe the knowledge of it. For this purpose an hypothesis must be made, and its justness tried by a strict comparison with facts. If they are found at variance, the assumed hypothesis must be relinquished with candour as erroneous, but should it, on the contrary prove, on a multitude of trials, invariably to accord with the results of observation, as nearly as our means of determination authorise us to expect, we are warranted in believing that the principle of nature is obtained, as we then have all the proofs of its being so, which men can have of the justness of their theories: a constant and perfect agreement with the phenomena, as far as can be discovered.

The great criterion in the present case is, whether on the conversion of a substance into its several compounds, and of these into one another, the simple ratios always obtain which the principles of the theory require. Amongst the multitude of instances which I could adduce, in support of such being the fact, I will, for the sake of brevity, confine myself to a few in the substances which have come under consideration above, as they will likewise give the grounds on which some of the proportions in the table have been assigned, and every chemist, by a careful repetition of the experiments, may easily determine for himself to what attention the present theory is entitled.

Lead   -   - $= \frac{3}{2}$ of sulphate of lead
                                     $= \frac{6}{5}$ of sulphuret of lead
Sulphuret of lead   - $= \frac{5}{6}$ of lead
                                     $= \frac{5}{4}$ of sulphate of lead
Sulphate of lead   - $= \frac{2}{3}$ of lead
                                     $= \frac{4}{5}$ of sulphuret of lead
Antimony   -   - $= \frac{4}{5}$ of powder of algoroth
                                     $= \frac{4}{5}$ of sulphuret of antimony
Sulphuret of antimony - $= \frac{10}{9}$ of powder of algoroth.

In the experiments by which these relations were ascertained, the portion of powder of algoroth and sulphate of lead dissolved in the precipitating and washing waters, was scrupulously collected.

The importance of a knowledge of the true quantity in which matters combine, is too evident to require to be dwelt upon; but this importance will be greatly augmented, if it should prove that this quantity is, as has been suggested, expressive of the forces with which they attract each other. It is perhaps in the form of matters that we shall find the cause of the proportions in which they unite, and a proof, *a priori*, of the system here maintained.

I have examined some of the grey ores of copper in tetraedral crystals; but the notes of my experiments are in England. I can, however, say, that they do contain antimony, and that they do not contain iron in any material quantity. With respect to the proportions of the constituent parts, I cannot now speak with any certainty; but, I think, that at least some species of fahlertz contain a smaller portion of sulphuret of antimony, than the fahlertz does which exists as an element in the foregoing compound one.

Of the Form of this Substance.

Of the seventeen figures which have been given, as of the crystals of this compound sulphuret, in Part II. of the volume of the Transactions for 1804, great part are acknowledged to have no existence, nor are indeed any of them consistent with nature.

This substance seems to have yet offered but one form, and which is represented in the annexed Plate under its two principal appearances; that is, having the primitive faces, the predominant ones of the prism; and having the secondary ones such, and which will be fully sufficient to make it known. In the first infancy of the study of crystals, it might be necessary to attend to every, the most trifling, variation of them, to trace each of their changes, step by step, to, as it were, spell the subject; but in the state to which the science has now attained, to continue to do so would be not only superfluous, but most truly puerile.

I have a very small, but very regular, crystal of the form of Fig. 1.

Fig. 1

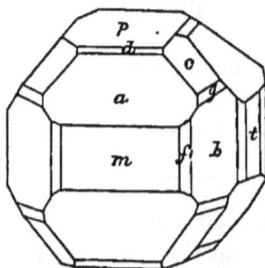

Fig. 2.

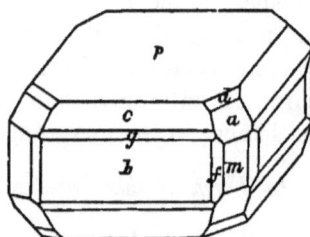

$$m\,p = 90°$$
$$m\,t = 90°$$
$$a\,m = 135°$$
$$m\,b = 135°$$
$$c\,b = 125°\ \ 15'\ \ 52''$$
$$g\,b = 144°\ \ 44'\ \ \ 8''$$
$$d\,m = 116°\ \ 33'\ \ 54''$$
$$f\,m = 153°\ \ 26'\ \ \ 6''$$

By mensuration the faces a and m appear to form together an angle of about 135°, and the faces c and b an angle of about 125°.

It is said in the account above quoted, that the primitive form of this matter is a rectangular tetraedral prism, but no proofs of this have been offered; nor have the dimensions of this prism been given, a circumstance of the first moment to the determination of true or primitive form, nor have any quantities been assigned to the decrements supposed. I will, therefore, supply these very important omissions.

That the atom of this substance is a rectangular tetraedral prism, is inferable, not from the striæ on the crystals, for striæ are by no means invariably indicative of a decrement in the direction of them ; but from the angles which the faces a and c make with the faces m and b, and these angles also prove, that the height of this prism is equal to the side of its base, that is, that it is a cube.

Hence the face a is produced by a decrease of one row of atoms along the edge of the cube, and the angle it forms with the face m is really of 135°.

The face c is produced by a decrease of two rows of atoms at the corners of the cube, and the angle it forms with the face b is $= 125°\ 15'\ 52''$.

The face b being produced like the face a, forms the same angle with the face m.

No crystal I possess, has enabled me to measure the inclinations of the faces g, d, or f; should the face g, as is presumable, result from a decrease of one row of atoms at the corners of the cube, it will form with the face b, an angle of 144° 44' 8'', and if the faces d and f are, as is also probable, produced by a decrease of two rows of atoms along the edges of the cube, the first will form an angle of 116° 33' 54'', and the latter one of 153° 26' 6'', with the face m.

The angles assigned here differ considerably from those given in the former account of these crystals; but the angles there given have not only appeared to me to be contradicted by observation, but, crystallographically considered, are inconsistent with each other, as the tetraedral prism of dimensions to produce an angle of 135° by a decrement along its edge, would not afford angles of 140° and 120° by decrements at its corners.

The sum of the faces of these crystals is 50.

ON THE COMPOSITION OF ZEOLITE.

From the Philosophical Transactions of the Royal Society of London,
Vol. CI, p. 171.—Read February 7, 1811.

MINERAL bodies being, in fact, *native chemical preparations*, perfectly analogous to those of the laboratory of art, it is only by chemical means, that their species can be ascertained with any degree of certainty, especially under all the variations of mechanical state and intimate admixture with each other, to which they are subject.

And accordingly, we see those methods which profess to supersede the necessity of chemistry in mineralogy, and to decide upon the species of it by other means than her's, yet bringing an unavoidable tribute of homage to her superior powers, by turning to her for a solution of the difficulties which continually arise to them, and to obtain firm grounds to relinquish or adopt the conclusions to which the principles they employ, lead them.

Zeolite and natrolite have been universally admitted to be species distinct from each other, from Mr. KLAPROTH having discovered a considerable quantity of soda and no lime, in the composition of the latter, while Mr. VAUQUELIN had not found any portion of either of the fixed alkalies, but a considerable one of lime, in his analysis of zeolite.*

The natrolite has been lately met with under a regular crystalline form, and this form appears to be perfectly similar to that of zeolite, but Mr. HAÜY has not judged himself warranted by this circumstance, to consider these two bodies as of the same species, because zeolite, he says, " does not contain an atom of soda."†

I had many years ago found soda in what I considered to

---

* Journal des Mines, No. XLIV.
† Journal des Mines, No. CL. Juin 1810, p. 458.

be zeolites, which I had collected in the island of Staffa, having formed GLAUBER's salt by treating them with sulphuric acid; and I have since repeatedly ascertained the presence of the same principle in similar stones from various other places; and Dr. HUTTON and Dr. KENNEDY, had likewise detected soda in bodies, to which they gave the name of zeolite.

There was, however, no certainty that the subjects of any of these experiments were of the same nature as what Mr. VAUQUELIN had examined, were of that species which Mr. HAÜY calls mesotype.

Mr. HAÜY was so obliging as to send me lately, some specimens of minerals. There happened to be amongst them a cluster of zeolite in rectangular tetrahedral prisms, terminated by obtuse tetrahedral pyramids whose faces coincided with those of the prism. These crystals were of a considerable size, and perfectly homogeneous, and labelled by himself " *Mesotype pyramidée du depart. de Puy de Dôme.*" I availed myself of this very favourable opportunity, to ascertain whether the mesotype of Mr. HAÜY and natrolite, did or did not differ in their composition, and the results of the experiments have been entirely unfavourable to their separation, as the following account of them will show.

10 grains of this zeolite being kept red hot for five minutes lost 0.75 grains, and became opaque and friable. In a second experiment, 10 grains being exposed for 10 minutes to a stronger fire, lost 0.95 grains, and consolidated into a hard transparent state.

10 grains of this zeolite, which had not been heated, were reduced to a fine powder, and diluted muriatic acid poured upon it. On standing some hours, without any application of heat, the zeolite entirely dissolved, and some hours after, the solution became a jelly : this jelly was evaporated to a dry state, and then made red hot.

Water was repeatedly poured on to this ignited matter till nothing more could be extracted from it. This solution was gently evaporated to a dry state, and this residuum made slightly red hot. It then weighed 3.15 grains. It was *muriate of soda.*

The solution of this muriate of soda being tried with solutions of carbonate of ammonia and oxalic acid, did not afford the least precipitate, which would have happened had the zeolite contained any lime, as the muriate of lime* would not have been decomposed by the ignition.

The remaining matter, from which this muriate of soda had been extracted, was repeatedly digested with marine acid, till all that was soluble was dissolved. What remained was silica, and, after being made red hot, weighed 4.9 grains.

The muriatic solution, which had been decanted off from the silica, was exhaled to a dry state, and the matter left made red hot. It was alumina.

To discover whether any magnesia was contained amongst this alumina, it was dissolved in sulphuric acid, the solution evaporated to a dry state, and ignited. Water did extract some saline matter from this ignited alumina, but it had not at all the appearance of sulphate of magnesia, and proved to be some sulphate of alumina which had escaped decomposition, for on an addition of sulphate of ammonia to it, it produced crystals of compound sulphate of alumina and ammonia, in regular octahedrons.

This alum and alumina were again mixed and digested in ammonia, and the whole dried and made red hot. The alumina left, weighed 3.1 grains.

Being suspected to contain still some sulphuric acid, this alumina was dissolved in nitric acid, and an excess of acetate of barytes added. A precipitate of sulphate of barytes fell, which after being edulcorated and made red hot, weighed 1.2 grains. If we admit ⅓ of sulphate of barytes to be sulphuric acid, the quantity of the alumina will be $= 3.1 - 0.4 = 2.7$ grains.

* These names are retained for the present, as being familiar, though, since Mr. DAVY's important discovery of the nature of what was called oxymuriatic acid, the substances to which they are applied, are known not to be salts, but metallic compounds analogous to oxides.

From the experiments of Dr. MARCET,* it appears that 3.15 grains of muriate of soda, afford 1.7 grains of soda.

Hence, according to the foregoing experiments, the 10 grains of zeolite analysed, consisted of

Silica	-	-	-	4.90
Alumina	-	-	-	2.70
Soda	-	-	-	1.70
Ice	-	-	-	0.95

$$10.25$$

As these experiments had been undertaken more for the purpose of ascertaining the nature of the component parts of this zeolite than their proportions, the object of them was considered as accomplished, although perfect accuracy in the latter respect, had not been attained, and which, indeed, the analysis we possess of natrolite by the illustrious chemist of Berlin, renders unnecessary.

I am induced to prefer the name of zeolite for this species of stone, to any other name, from an unwillingness to obliterate entirely from the nomenclature of mineralogy, while arbitrary names are retained in it, all trace of one of the discoveries of the greatest mineralogist who has yet appeared, and which, at the time it was made, was considered as, and was, a very considerable one, being the first addition of an earthy species, made by scientific means, to those established immemorially by miners and lapidaries, and hence having, with tungstein and nickel, led the way to the great and brilliant extension which mineralogy has since received. And, of the several substances, which, from the state of science in his time, certain common qualities induced Baron CRONSTEDT to associate together under the name of zeolite; it is this which has been most immediately understood as such, and whose qualities have been assumed as the characteristic ones of the species.

Indeed, I think that the name imposed on a substance by

---

* Phil. Trans. 1807.

the discoverer of it, ought to be held in some degree sacred, and not altered without the most urgent necessity for doing it. It is but a feeble and just retribution of respect for the service which he has rendered to science.

Professor STRUVE, of Lausanne, whose skill in mineralogy is well known, having mentioned to me, in one of his letters, that from some experiments of his own, he was led to suspect the existence of phosphoric acid in several stones, and particularly in the zeolite of Auvergne, I have directed my enquiries to this point, but have not found the phosphoric, or any other acknowledged mineral acid, in this zeolite.

Many persons, from experiencing much difficulty in comprehending the combination together of the earths, have been led to suppose the existence of undiscovered acids in stony crystals. If quartz be itself considered as an acid, to which order of bodies its qualities much more nearly assimilate it, than to the earths, their composition becomes readily intelligible. They will then be neutral salts, silicates, either simple or compound. Zeolite will be a compound salt, a hydrated silicate of alumina and soda, and hence a compound of alumina not very dissimilar to alum. And topaz, whose singular ingredients, discovered by Mr. KLAPROTH, have called forth a query from the celebrated Mr. VAUQUE-LIN, with regard to the mode of their existence together,[*] will be likewise a compound salt, consisting of silicate of alumina, and fluate of alumina.

Our acquaintance with the composition of the several mineral substances, is yet far too inaccurate to render it possible to point out with any degree of certainty, the one of which zeolite is an hydrate, however the agreement of the two substances in the nature of their constituent parts, and in their being both electrical by heat, directs conjecture towards tourmaline.

St. James's Place, Jan. 22, 1811.

---

* Annales du Museum d'Hist. Nat. tome 6, p. 24.

ON A SUBSTANCE FROM THE ELM TREE, CALLED ULMIN.

From the Philosophical Transactions of the Royal Society of London, Vol. CIII, Part I, 1813, p. 64.—Read December 10, 1812.

1. The substance now denominated Ulmin was first made known by the celebrated Mr. KLAPROTH, to whom nearly every department of chemistry is under numerous and great obligations.*

Ulmin has been ranked by Dr. THOMSON, in his System of Chemistry, as a distinct vegetable principle, on the ground of its possessing qualities totally peculiar and extraordinary. It is said, that though in its original state easily soluble in water and wholly insoluble in alcohol and ether, it changes, when nitric, or oxymuriatic acid is poured into its solution, into a resinous substance no longer soluble in water, but soluble in alcohol, and this singular alteration is attributed to the union to it of a small portion of oxygen which it has acquired from these acids.* Being possessed of some of this substance which had been sent to me some years ago from Palermo, by the same person from whom Mr. KLAPROTH had received it, I became induced, by the foregoing account, to pay attention to it, and have observed facts which appear to warrant a different etiology of its phenomena, and opinion of its nature, from what has been given of them.

The ulmin made use of in the following experiments, had been freed from the fragments of bark by solution in water and filtration, and recovered in a dry state by the evaporation of the solution on a water bath.

2. In lumps, ulmin appears black, but in thin pieces it is seen to be transparent, and of a deep red colour.

---

* Dr. THOMSON's Syst. of Chem. Vol. IV, p. 696. Fourth edition.

In a dilute state, solution of ulmin is yellow ; in a con-centrated one, dark red, and not unlike blood.

When solution of ulmin dries, either spontaneously or by being heated, the ulmin divides into long narrow strips dis-posed in rays to the centre, which curl up and detach them-selves from the vessel, and the fluid part seems to draw together, and becomes remarkably protuberant. Solution of ulmin slowly and feebly restores the colour of turnsol paper reddened by an acid.

3. Dilute nitric acid being poured into a solution of ulmin, a copious precipitate immediately formed. The mixture was thrown on a filter. The matter which has been considered as a resin remained on the paper, and a clear yellow liquor came through. This yellow solution, on evaporation, produced a number of prismatic crystals look-ing like nitrate of potash. They were tinged yellow by some of the resin. This mixture, heated in a gold dish, deflagrated with violence, and a large quantity of fixed alkali remained.

Dilute muriatic acid caused an exactly similar precipita-tion in solution of ulmin to nitric acid, and the precipitate was the same resin-like substance. The filtered liquor afforded a quantity of saline matter, which, after being freed by ignition from a portion of dissolved resin, shot into pure white cubes of muriate of potash, as appeared by decomposing them by nitric acid.

Sulphuric, phosphoric, oxalic, tartaric, and citric acids, occasioned a similar precipitation in solution of ulmin.

Distilled vinegar produced no turbidness in it; and the mixture being exhaled to dryness, at a gentle heat, was found to be again wholly soluble in water. But when the mixture was made to boil, some decomposition took place. On adding muriatic acid to a mixture of solution of ulmin and distilled vinegar, a precipitate was produced, as in a mere solution in water.

The nitric and muriatic acids received a small quantity of lime and iron from the ulmin, and I believe also a little magnesia; but these can be considered only as foreign admixtures.

4. To acquire an idea of the quantity of potash in ulmin, 4 grains of ulmin were decomposed by nitric acid. They afforded 2.4 grains of resin-like matter. The nitrate of potash obtained was heated to deflagration, in small quantities at a time, in a platina crucible to free it from resin. The alkali produced was supersaturated with nitric acid, dried, and slightly fused. It then weighed 1.2 grains. If we admit $\frac{1}{2}$ of nitrate of potash to be alkali, this will denote $\frac{15}{100}$ of potash in ulmin.

5 grains of ulmin were decomposed by muriatic acid. The resinous matter weighed 3.3 grains, and the muriate of potash, after being ignited, dissolved away from the charcoal, dried, and again made red hot, weighed 1.4 grains. If we suppose $\frac{2}{3}$ of muriate of potash to be alkali, this will indicate $\frac{19}{100}$ of potash in ulmin.

2 grains of ulmin were made red hot in a gold crucible. It then weighed only 1.05 grain. The form of the flakes was in no degree altered, but they had acquired the blue and yellow colours of heated steel, of which they had likewise the metallic aspect and lustre, and could difficultly, if at all, have been distinguished by the eye from heated steel-filings, or fragments of slender watch-springs. Water immediately destroyed their metallic appearance.

Muriatic acid, poured on, caused a strong effervescence, and formed muriate of potash, which, freed from all charcoal, and made red hot, weighed 0.6 grain, corresponding to $\frac{20}{100}$ of potash in ulmin.

These experiments assign about $\frac{1}{5}$ for the quantity of potash in ulmin, but as it is impossible to operate, on so small a scale, on such substances without loss, it is probable that it even exceeds this proportion.

5. The substance separated from ulmin by acids has the following qualities:

It is very glossy, and has a resinous appearance.

4

In lumps it appears black, but in minute fragments it is found to be transparent, and of a garnet-red colour.

It burns with flame, and is reduced to white ashes.

Alcohol dissolves it, but only in very small quantity.

Water likewise dissolves it, but also only in very small quantity. Acids cause a precipitate in this solution, though this resin-like matter appears neither to contain any alkali, nor to retain any of the acid by means of which it was obtained.

Its solution in water seems to redden turnsol paper.

Neither ammonia, nor carbonate of soda, promote its solution in cold water.

On adding a small quantity of potash to water in which it lies, it dissolves immediately and abundantly. This solution has all the qualities of a solution of ulmin, and, on exhalation, leaves a matter precisely like it, which cracks and separates from the glass, and does not grow moist in the air, &c.

Hence it appears that ulmin is not a simple vegetable principle of anomalous qualities, but a combination with potash of a red, or more properly a high yellow matter, which, if not of a peculiar genus, seems rather more related to the extractives than to the resins.

English Ulmin.

I collected, from an elm tree in Kensington gardens, a small quantity of a black shining substance which looked like ulmin.

It was readily soluble in water, and the solution was in colour and appearance exactly similar to a solution of ulmin.

This solution, exhaled to a dry state on a water-bath, left a matter exactly like ulmin, and which cracked and divided as ulmin does, when dried in the same manner. It did not, however, rise up from the watch-glass in long strips, like the Sicilian kind, but this may have been owing partly to its small quantity, which occasioned it to be spread very thin on the watch-glass, and partly to its containing a considerable excess of alkali, for it differed also from the Palermo ulmin by becoming soft in the air, and its solution strongly restored the blue colour of reddened turnsol paper.

Nitric acid, added to a filtered solution of this ulmin, immediately caused a precipitate in it, and the filtered solution, on evaporation, afforded numerous crystals of nitrate of potash.

This English ulmin made a considerable effervescence with acetous acid, which the Palermo ulmin had not been observed to do. This acetous solution, in which the acid was in excess, was exhaled dry, and repeatedly washed with spirit of wine. No part of the brown matter dissolved. Water dissolved this brown residuum readily and entirely. This solution did not sensibly restore the blue colour of reddened turnsol paper. Exhaled to a dry state, the matter left did not separate from the watch-glass quite as freely as Palermo ulmin, which had been treated with acetous acid; but it seemed no longer to grow moist in the air. Redissolved in water, and nitric acid added, the mixture became thick from a copious precipitate.

The spirit of wine contained a quantity of acetate of potash.

The excess of alkali, in this English ulmin, may be owing to the tree from which it was collected having been affected with the disease, which produces the alkaline ulcer to which the elm is subject.

Sap of the Elm Tree.

Thinking that the production of ulmin by the plant might not be the consequence of disease, and that it might exist in the healthy sap, a bit of elm twig, gathered in the beginning of last July, was cut into thin slices and boiled in water. It afforded a brown solution, like a solution of ulmin. Exhaled to dryness, this solution left a dark brown substance, in appearance similar to ulmin, but on adding water to this dry mass, a large quantity of brown glutinous matter remained insoluble. The mixture being thrown on a filter, a clear yellow liquor passed, which may have contained ulmin, but the quantity was too small to admit of satisfactory conclusions.

Perhaps older wood, the juice of which was more per-fected, would afford other results, since ulmin appears to be the product of old trees; but the inquiry, being merely collateral to the object 1 had originally in view, was not persevered in.

ON A SALINE SUBSTANCE FROM MOUNT VESUVIUS.

From the Philosophical Transactions of the Royal Society of London, Vol. CIII, Part I, 1813, p. 256.—Read July 8, 1813.

It has very long appeared to me, that when the earth is considered with attention, innumerable circumstances are perceived, which cannot but lead to the belief, that it has once been in a state of general conflagration. The exist-ence in the skies of planetary bodies, which seem to be actually burning, and the appearances of original fire dis-cernible on our globe, I have conceived to be mutually cor-roborative of each other; and at the time when no answers could be given to the most essential objections to the hypothesis, the mass of facts in favour of it fully justified, I thought, the inference that our habitation is an extinct comet or sun.

The mighty difficulties which formerly assailed this opinion, great modern discoveries have dissipated. Ac-quainted now, that the bases of alkalies and earths are metals, eminently oxydable, we are no longer embarrassed either for the pabulum of the inflammation, or to account for the products of it.

In the primitive strata, we behold the result of the combustion. In them we see the oxyd collected on the surface of the calcining mass, first melted by the heat, then by its increase arresting farther combination, and extinguishing the fires which had generated it, and in fine become solid and crystallized over the metallic ball.

Every thing tells that a large body of combustible matter still remains enclosed within this stony envelope, and of which volcanic eruptions are partial and small accensions.

Under this point of view, an high interest attaches itself to volcanoes, and their ejections. They cease to be local phenomena; they become principal elements in the history of our globe; they connect its present with its former condition; and we have good grounds for supposing, that in their flames are to be read its future destinies.

In support of the igneous origin, here attributed to the primitive strata, I will observe, that not only no crystal imbedded in them, such as quartz, garnet, tourmaline, &c. has ever been seen enclosing drops of water; but that none of the materials of these strata contain water in any state.

a. The present saline substance was sent to me from Naples to Florence, where I was, in May 1794, with a request to ascertain its nature. The general examination which I then made of it, shewed it to be principally what was at that time called *vitriolated tartar*, and it was in consequence mentioned as such in an Italian publication soon after. But as this denomination, surprising at that period, was not supported by the relation of any experiments, or the citation of any authority, no attention was paid to it; and the existence of this species of salt, native in the earth, has not been admitted by mineralogists, no mention being made of it, I believe, in any mineralogical work published since.

b. I was informed by letter, that it had " flowed out liquid from a small aperture in the cone of Vesuvius," and which I apprehend to have happened in 1792 or 1793.

c. The masses of this salt are perfectly irregular, their texture compact, their colour a clouded mixture of white, of the green of copper, and of a rusty yellow, and in some places are specks and streaks of black.

d. A fragment melted on the charcoal at the blow-pipe formed hepar sulphuris.

e. A piece weighing 9.5 grains was so strongly heated in a platina crucible, that it melted and flowed level over the bottom of it, but did not lose the least weight.

f. Not the slightest fume could be perceived on holding a glass tube wetted with marine acid over some of this salt, while triturating in a mortar with liquid potash; but a similar mixture being made in a bottle, and which was immediately closed with a cork, to which was fixed a bit of reddened litmus paper, the blue colour of the paper was restored.

g. Being dissolved in water, there was a small sandy residue, which consisted of green particles of a cupreous nature, of a yellow ochraceous powder, and of minute crystals of a metallic aspect of red oxyd of iron, by which the black spots in the mass had been occasioned.* Mr. KLAPROTH found a similar admixture in muriate of soda from Vesuvius.†

h. The solution had a feeble green tint. It did not alter blue or reddened turnsol paper.

i. Prussiate of soda-and-iron threw down a small quantity of red prussiate of copper from it. Liver of sulphur and tincture of galls likewise caused very small precipitations.

j. Carbonate of soda, and oxalate of potash, and solutions

---

* What mineralogists denominate speculary iron ore, *Fer oligiste* of Mr. HAÜY, appears to be merely red oxyd of iron in crystals; red hematite the same substance in the state of stalactite; and red ochres the same in a pulverulent form. The hematites which afford a yellow powder are hydrates of iron.

† Essays, Vol. II. p. 67, Eng. Trans.

of magnesia, clay, copper, iron, and zinc, either had no effects, or extremely slight ones.

k. Solution of sulphate of silver produced a white curd-like precipitate. 9.35 grains of this salt (the weight of the insoluble matter being deducted) afforded 1.05 grains of slightly melted muriate, or chloride, of silver. This precipitate was equally produced after the salt had been made strongly red hot, so that it was not owing to a portion of sal ammoniac.

l. Tartaric acid, and muriate of platinum, occasioned the precipitates in its solution which indicate potash.

m. Nitrate of lime did not form any immediate precipitate in a dilute solution of it; but in a short time, numerous minute prismatic crystals of hydrate of sulphate of lime were generated.

n. Nitrate of barytes poured into a solution containing 9.8 grains of this salt afforded a precipitate, which after being ignited weighed 12.3 grains. The filtered solution crystallized entirely into nitrate of potash mixed with a few rhomboides of nitrate of soda.

o. Some of this salt finely pulverized was treated with alcohol. This alcohol on exhaling left a number of minute cubic crystals, which proved, by the test of nitric acid, to be muriate of soda. Prussiate of soda-and-iron caused a red precipitate of prussiate of copper in this alcoholic solution.

p. The solution of this salt afforded, by crystallization, sulphate of potash in its usual forms, and some prismatic crystals of hydrate of sulphate of soda.

q. To discover what had occasioned the precipitate with galls, (*i*) since copper has not this quality, a portion of this salt, which had been recovered by evaporation from a filtered solution of it, was made red hot in a platina crucible. On extraction of the saline part by water, a very small quantity of a black powder was obtained. Ammonia dissolved only part of it, which was copper. The rest being digested with muriatic acid, and prussiate of soda-and-iron added, a fine Prussian blue was formed.

r. From several of the foregoing experiments, it appeared that no sensible quantity of any of the mineral acids, besides the sulphuric and muriatic, existed in combination with alkali in this volcanic salt. But Mr. TENNANT, whose many and highly important discoveries have so greatly contributed to the progress of chemical science, having detected disengaged boracic acid amongst the volcanic productions of the Lipari islands, and suggested that it might be a more general product of volcanoes than had been suspected,* it became important to ascertain whether the presence of any in this salt proved Vesuvius likewise to be a source of this acid. Alcohol heated on a portion of it in fine powder, and then burned on it, did not however shew the least green hue in its flame.

s. To ascertain the proportions of the ingredients of this saline substance, the following experiments were made:

10 grains of sulphate of potash of the shops were dissolved in 200 grains of water, and an excess of muriate of platina added. The precipitate edulcorated with 100 grains of water, and dried on a water bath, weighed 24.1 grains.

10 grains of the saline part of the native salt, treated precisely in every respect in the same way, afforded 17.2 grains of precipitated muriate of platina-and-potash.

If 24.1 grains of this precipitate correspond to 10 grains of sulphate of potash, 17.2 grains of it correspond to 7.14 grains of this salt.

It has been seen (*n*) that 10 grains of the saline part of this volcanic salt would have afforded 12.55 grains of sulphate of barytes.

But 7.14 grains of sulphate of potash form only 9.42 grains of sulphate of barytes,† and therefore the remaining 3.13 grains of sulphate of barytes would be produced by the

---

* Trans. of the Geolog. Soc.
† Dr. MARCET on Dropsical Fluids.

sulphate of soda, and correspond to 1.86 grains of it in an
arid state, or uncombined with ice.*

10 grains of the saline part of this native salt would have
produced 1.12 grains of ignited muriate of silver (k).   By
accurate experiments 241 grains of ignited muriate of silver
have been found to correspond to 100 grains of ignited muriate of soda.†

Consequently the soluble portion of the present Vesuvian
salt consists of

Sulphate of potash	- -	7.14
Sulphate of soda	- -	1.86
Muriate of soda	- -	0.46
Muriate of ammonia		
Muriate of copper	- - -	0.54
Muriate of iron		

10.00

t. The insoluble sandy residue (g) having been thoroughly
edulcorated, dilute nitric acid was put to it.   A green solution formed without any effervescence.   Acetate of barytes
scarcely rendered this solution turbid; but nitrate of silver
produced a copious curd-like precipitate, and iron abundantly threw down copper from it.   The green grains enclosed
in this native sulphate of potash, appear, therefore, to be a
submuriate of copper, of the same species as that of the
green sands of Peru and Chili.

Muriatic acid dissolved the yellow ochraceous powder,
and prussiate of soda-and-iron produced Prussian blue.   I
am inclined to believe this yellow powder to be a submuriate of iron, but its small quantity, and the admixture of
the submuriate of copper, were impediments to entirely
satisfactory results.   Such a submuriate of iron, though, if
I mistake not, overlooked by chemists, exists, for the precipitate which oxygen occasions in solution of green muriate
of iron, contains marine acid.

---

* Prof. KLAPROTH's Essays, Vol. 1, p. 282.
† Dr. HENRY, Phil. Trans. 1810.

Possibly this yellow powder, and the crystals of speculary iron which exist in this Vesuvian salt, have been produced by a natural sublimation of muriate of iron, similar to that of the experiment of the Duke d'AYEN, recorded by MAC-QUER,* and which was known long before to Mr. BOYLE and Dr. LEWIS.†

This Vesuvian salt, considered in its totality, has presented no less than nine distinct species of matters, and a more rigorous investigation, than I was willing to bestow on it, would probably add to their number.

July 3, 1813.

A FEW FACTS RELATIVE TO THE COLOURING MATTERS OF SOME VEGETABLES.

From the Philosophical Transactions of the Royal Society of London, Vol. CVIII, p. 110.—Read December 18, 1817.

I BEGAN, a great many years ago, some researches on the colouring matters of vegetables. From the enquiry being to be prosecuted only at a particular season of the year, the great delicacy of the experiments, and the great care required in them, and consequently the trouble with which they were attended, very little was done. I have now no idea of pursuing the subject.

In destroying lately the memorandums of the experiments which had been made, a few scattered facts were met with which seemed deserving of being preserved. They are here offered, in hopes that they will induce some other person to give extension to an investigation interesting to chemistry and to the art of dying.

---

* *Dict. de Chimie*, Art. *Fer.*
† A course of practical chemistry by WILLIAM LEWIS, 1746, p. 63, note *f.*

Turnsol.

M. FOURCROY has advanced, somewhere, that turnsol is essentially of a red colour; and that it is made blue by an addition of carbonate of soda to it; and he says that he has extracted this salt from the turnsol of the shops.

If turnsol contained carbonate of soda, its infusions should precipitate earths and metals from acids.

I did not find an infusion of turnsol in water to have the least effect on solutions of muriate of lime, nitrate of lead, muriate of platina, or oxalate of potash.

Its tinctures, or infusions, consequently, contain neither any alkali, nor any lime; nor probably any acid, either loose or combined. This is unfavourable to the opinion of urine being employed in the preparation of turnsol.

I put a little sulphuric acid into a tincture of turnsol, then added chalk, and heated; and the blue colour was restored. It appears, therefore, that the natural colour of turnsol is not red, but blue, since it is such when neither disengaged acid or alkali is present.

No addition of chalk brought the cold liquor back to a blue colour; the carbonic acid absorbed by it, during the effervescence of the carbonate of lime, being sufficient to keep it red.

Some turnsol was put into distilled vinegar. An effervescence arose; and after some time the acid was become neutralized. On examining the mixture with a glass, there were seen, at the bottom of the vessel, a multitude of grains like sand. It was found on trial that these grains were carbonate of lime; probably of slightly calcined Carrara marble.

When turnsol is treated with water till this no longer acquires any color whatever, the remaining insoluble matter is nearly as blue as at first.

Acids made this blue insoluble matter red, but did not extract any red tincture.

Carbonate of soda did not affect it.

If the vegetable part of this blue residuum is burned away, or it is washed off with water, a portion of smalt is obtained.

On exhaling, on a water bath, a tincture of turnsol, the colouring matter is left in a dry state.

This matter heated in a platina spoon over a candle, tumefied considerably, as much as starch does, became black and smoked, but did not readily inflame, nor did it burn away till the blow pipe was applied. It then burned pretty readily, leaving a large quantity of white saline matter. This saline matter saturated by nitric acid afforded crystals of nitrate of potash, and some minute crystals like hydrous sulphate of lime.

Is this potash merely that portion of this matter which exists in all vegetable substances? or is the colouring matter of turnsol a compound, analogous to ulmin, of a vegetable principle and potash? Its low combustibility gives some sanction to this idea.

Of the colouring matter of the violet.

The violet is well known to be coloured by a blue matter which acids change to red; and alkalies and their carbonates first to green and then to yellow.

This same matter is the tinging principle of many other vegetables: of some, in its blue state; of others, made red by an acid.

If the petals of the red rose are triturated with a little water and carbonate of lime, a blue liquor is obtained. Alkalis, and soluble carbonates of alkalis, render this blue liquor green; and acids restore its red colour.

The colouring matter of the violet exists in the petals of red clover, the red tips of those of the common daisy of the fields, of the blue hyacinth, the holly hock, lavender, in the inner leaves of the artichoke, and in numerous other flowers. It likewise, made red by an acid, colours the skin of several plumbs, and, I think, of the scarlet geranium, and of the pomegranate tree.

The red cabbage, and the rind of the long radish are also coloured by this principle. It is remarkable that these, on being merely bruised, become blue; and give a blue infusion with water. It is probable that the reddening acid in these cases is the carbonic; and which, on the rupture of the vessels which enclose it, escapes into the atmosphere.

Of sugar-loaf paper. •

This paper has been employed by BERGMAN as a chemical instrument. I am ignorant of what it is coloured with.

Sulphuric, muriatic, nitric, phosphoric, and oxalic acids make it red. Tartaric and citric acids, made rather yellow spots than red ones. Distilled vinegar, and acid of amber, had no affect on it.

Carbonate of soda and caustic potash did not alter the blue colour of this paper.

Water boiled on this paper acquired a vinous red colour; carbonate of lime put into this red liquor, did not affect its colour: nor did carbonate of soda or caustic potash change it to blue or green.

Cold dilute sulphuric acid extracted a strong yellow tincture from this boiled paper: carbonate of lime put to this yellow tincture made it blue; but on filtering, the liquor which passed was of a dirty greenish colour; and sulphuric acid did not make it red: a blue matter was left on the filter, which was not made red by acetous acid; but was so by sulphuric.

After this treatment the paper remained brown; seemingly such as it was before being dyed blue.

It should seem that there are at least two colouring matters in this paper; one red, which is extricable from it by water; the other blue, which requires the agency of an acid to extract it.

Its insolubility in water, and low degree of sensibility to acids, distinguish the blue matter from turnsol; to which its not being affected by alkalis otherwise much approximate it. Its easy solubility in dilute suphuric acid, and being reddened by it and several other acids, show it not to be indigo.

Of the black mulberry.

The expressed juice of this fruit is of a fine red colour.

Caustic potash made it green, which gradually became yellow.

Carbonate of soda did not make it green, but only blue.

Carbonate of ammonia changed it to a vinous red, rather than to blue; and this redness increased on standing.

Caustic ammonia made it bluer than its carbonate; but, on standing, the mixture became of the same vinous red.

The mulberry juice mixed with carbonate of lime became purple. On filtering, a red liquor passed; and the carbonate of lime left on the filter was blue. An addition of whitening to the red filtered liquor did not alter its colour; nor did this second portion of whitening become blue. Heating did not affect the red colour of this liquor; so that it was not owing to carbonic acid, disengaged from the carbonate of lime. Caustic potash instantly made this red liquor a fine green, and gradually yellow.

Sulphuric acid rendered all the above mixtures florid red. It is remarkable that the mixtures with ammonia, and carbonate of ammonia, which were become quite vinous red by standing, were made a perfect blue by the sulphuric acid before they were reddened by it. It would hence seem that the red colour, caused by these alkalis, was owing to an excess of them; and that in a less quantity they would have produced a blue.

The filter, into which the mixture of mulberry juice and chalk had been thrown, was become tinged blue. Water did not remove this colour. Sulphuric acid made this paper florid red. Caustic potash did not alter its blue colour; but put on the places made red by sulphuric acid, it restored the blue colour, but did not produce green.

Future experiments must decide whether this blue matter

is the same as that of turnsol; or as the blue matter which the experiments above have indicated in sugar-loaf paper.

The juices of many other fruits, as black cherries, red currants, the skin of the berries of the buckthorn, elder berries, privet berries, &c., seemed to be made only blue by mild fixed alkalis, but green by caustic. Puzzling anomalies, however, occasionally present themselves, which seem to show a near relation between the several blue colouring matters of vegetables, and their easy transition into one another.

The corn poppy.

The petals of the common red poppy of the fields rubbed on paper stain it of a reddish purple colour.

Solution of carbonate of soda put to this stain occasioned but little change in it.

Caustic potash made it green.

Caustic ammonia seemed not to have more effect on it than carbonate of soda.

Some poppy petals being bruised in a mixture of water and marine acid, formed a florid red solution: a superabundance of chalk added to this red liquor, did not make it blue; but turned it to a dark red colour exactly like port wine.

Some poppy petals bruised in a weak solution of carbonate of soda, and the mixture filtered, the liquor which came through was not at all blue, but of a dark red colour like port wine. Caustic potash made this red liquor green, which finally became yellow.

Some dried poppy petals of the shops, gave a strong obscure vinous tincture to cold water. This red tincture heated with whitening, did not alter to blue, but preserved its red colour.

These very imperfect experiments may perhaps suggest the idea, that the colouring matter of this flower is the same as the red colouring matter of the mulberry.

Of sap green.

The inspissated juice of the ripe, or semi-ripe, berries of the buckthorn, constitute the pigment called sap green; by the French, vert de vessie. This species of green matter is entirely different from the common green matter of vegetables.

It is soluble in water.

Carbonate of soda and caustic potash changed the solution of sap green to yellow. Paper tinged by sap green is a sensible test of alkalis.

Sulphuric, nitric, and marine acid, made it red. Carbonate of lime added to a reddened solution, restored the green colour, which therefore appears to be the proper colour of the substance.

The green colour, which the last infusions of galls present, appears to be different, both from the usual green of vegetables, and from sap green.

Some animal greens.

A green puceron, or aphis, being crushed on white paper, emitted a green juice, which was immediately made yellow by carbonate of potash (wrongly called sub-carbonate.)

There are small gnats of a green colour: crushed on paper, they make a green stain, which is permanent. Neither muriatic acid nor carbonate of soda altered this green colour. It is consequently of a different nature from the foregoing.

ON A NATIVE COMPOUND OF SULPHURET OF LEAD AND ARSENIC.

From Thomson's Annals of Philosophy, Vol. XIV., 1819, p. 96.

PARIS, *May* 19, 1819.

This mineral is found in Upper Valais, in Switzerland. It is lodged in a white, granose, compound carbonate of lime and magnesia. It is accompanied in this rock by regular crystals of yellow sulphuret of iron; by red sulphuret of arsenic; and by some other substances.

This compound sulphuret has a metallic aspect. It is of a grey colour; it is exceedingly brittle and soft; its fracture in some directions is perfectly vitreous; but in at least one direction, it is evidently tabular; but the size of the fragments I had, not exceeding coarse sand, precluded research with respect to crystalline construction. By trituration, this ore afforded a red powder.

At the blow-pipe, this ore melted instantly on the contact of the point of the flame. It smoked considerably; and a small flame was visible on the surface of the melted button. On cooling, this button forced out a quantity of fluid matter from its interior. During the fusion, the bead occasionally swelled up, and puffs of dense smoke issued from it; due evidently to a volatile matter, which the fire expelled from another less volatile. Finally, a button of a more fixed, less fusible, white metallic matter, extensible under the hammer, was left, and which proved to be lead.

Some bits of this compound sulphuret heated in a tube over a candle, melted, and a red sublimate rose, which became yellow on cooling, and looked like orpiment.

Some of this ore, being fused with nitre, deflagrated, and became a white oxide. The solution of this nitre afforded a white precipitate with muriate of barytes; and with

5

nitrate of silver, a brick-red precipitate of arseniate of silver.

The white precipitate by muriate of barytes was only partially soluble in nitric acid. The insoluble part of this precipitate, of which the quantity was so minute that no balance would have been sensible to it, was carefully collected on to a very small bit of charcoal fixed to a pin. It was then strongly heated at the blow-pipe. This bit of charcoal now put into a drop of water, placed on a silver coin, immediately made a black stain of sulphuret of silver on the coin. This is the nicest test I am acquainted with of the presence of sulphur, or sulphuric acid, in bodies.

The quantity I possessed of this mineral for experiment was very small. The above trials were made with particles little more than visible; however, the results, I think, sufficiently establish the nature of the constituent parts. Their respective proportions must remain for inquiries on another scale.

From Thomson's Annals of Philosophy, Vol. XVI, 1820, p. 100.

Compound of Sulphuret of Lead and Arsenic.—This is a new mineral species discovered by Mr. Smithson, and described by him in the *Annals of Philosophy*, xiv. 96. It was found in a magnesian lime rock in the Upper Valais. It has a metallic aspect, a grey colour, and a fracture in some directions vitreous, in others foliated. When triturated, yields a red powder. Mr. Smithson, by a set of very minute but satisfactory experiments, demonstrated that its constituents were sulphur, arsenic, and lead.

ON NATIVE HYDROUS ALUMINATE OF LEAD, OR PLOMB GOMME.

From Thomson's Annals of Philosophy, Vol. XIV, 1819, p. 31.

PARIS, *May* 22, 1819.

I see in the *Annals of Philosophy* for this month, which I have very lately received, an analysis by M. Berzelius of the mineral which was formerly known here under the name of "plomb gomme."

The first discovery of the composition of this singular substance belongs, however, to my illustrious and unfortunate friend, and indeed distant relative, the late Smithson Tennant. He ascertained when last at Paris, on pieces furnished him by M. Gillet de Laumont, that it was a combination of oxide of lead, alumina, and water.

At that time I received a small specimen of this rare ore from M. de Laumont, accompanied with a label, of which the following is a copy :

"Hydrate d'alumine et de plomb reconnu par Mr. Tennant, du Huelgoât, près Poullaouen, en Bretange (Finisterre) qui paroit etre la même substance decrite par Romé de l'Isle, tom. iii. de la Cristallographie, p. 399, comme plomb rouge en stalactite.

"J'en ai dit quelques mots en Mai, 1786, dans le Journal de Physique, p. 385, F. 16."

This ore is of a yellow colour ; it otherwise bears so great a resemblance to the siliceous substance found near Frankfort on the Mein, called Müllen glass, that it might be mistaken for it.

Suddenly heated, it decrepitated violently ; but heated slowly, it became white and opaque. The utmost fire did not appear to fuse it, or produce any further alteration in it.

It dissolved readily in borax into a colourless transparent glass, but no reduction of lead took place. Not having any carbonate of soda at hand, I added a particle of nitre, whose deflagration producing potash, lead was revived.

A bit, which had been made white by ignition, being wetted with nitrate of cobalt and again ignited, became blue.

Heated in a glass tube over a candle, it decrepitated, became opaque and white, and water sublimed.

Mr. Tennant mentioned to me a sort of explosion occasioned by the sudden expulsion of the water, and characteristic of this ore, which took place when it was heated at the blow-pipe. With the very minute particles I have tried, no effect of this sort was perceived.

The above characters will prove sufficient, I apprehend, to make this substance known when met with.

From Thomson's Annals of Philosophy, Vol. XVI, 1820, p. 100.

Plomb Gomme.—Mr. Smithson has given us some interesting details respecting the history and properties of this mineral, which is a *hydrous aluminate of lead.* It has a yellow colour, and is exceedingly similar in appearance to Mullen glass. When heated, it decrepitates violently; and if it be heated by the blow-pipe, in contact with an alkali, lead is reduced. Its nature was first ascertained by Mr. Tennant. Berzelius has lately analyzed it. The result of his analysis will be found in the *Annals of Philosophy,* xiii. 881. (See *Annals of Philosophy,* xiv. 31.)

ON A FIBROUS METALLIC COPPER.

From Thomson's Annals of Philosophy, Vol. XVI, 1820, p. 46.

PARIS, *March* 17, 1820.

SIR: There occur, in mineral collections, pieces of a copper slag, having fibres of metallic copper in its cavities. I have seen this fibrous copper erroneously placed among native coppers.

I possess samples of this kind from a foundery in the Hartz. The metallic copper in the cavities, or air-holes, is so delicately slender as to be a metallic wool.

From several considerations, it appeared to me to be beyond all doubt that the opinion of these fibres having been produced by crystallization was perfectly inadmissible; and I was for a very long time totally unable to come to any conjecture with respect to the mode in which they had originated.

Looking on one of these specimens this morning, an idea struck me which is, I am convinced, the solution of this knotty problem.

It occurred to me that these fibres had been generated at the instant of consolidation of the fused slag. That by its shrinking at that moment, it had compressed drops of copper, still in a fluid state, dispersed in its substance, and squeezed a portion of it through the minute spaces between its particles, under this fibrous form, into its cavities, or airholes.

For this operation to take place, the concurrence of several conditions is required. The slag must be so thick and pasty as to retain metallic copper scattered through it. It must have developed bubbles of some gas which have occasioned vacuities in it. It must be less fusible than the copper, but in so very small a degree that the copper consolidates as the fibres of it are formed.

It is evident that on this supposition these fibres of copper are produced by a process entirely the same as that employed for the manufactory of macaroni and vermicelli; and which are made by forcing paste through small apertures by the pressure of a syringe. It is wire-drawing performed inversely—by propulsion instead of traction.

As soon as this hypothesis had presented itself to me, I became anxious to ascertain whether I could give birth to this fibrous copper at the blow-pipe. I melted a small fragment of the slag; and, on breaking it, I had the gratification of finding its little cavities lined with minute fibres of metallic copper as those of its greater prototype.

I wished now to form the slag itself which was to afford the copper fibres. As I had ascertained the slag of the

Hartz to consist of sulphur, copper, and iron, I had recourse to the yellow sulphuret of copper and iron. To produce the required portion of metallic copper, I calcined some small fragments of this yellow ore at the tip of the exterior flame. Finding that I had exceeded the proper point, and rendered them too infusible, I added a little of the raw ore; and after encountering a few difficulties succeeded in producing a little mass of slag, whose internal cavities presented me, on breaking it, with the fibres of copper which were the object of my toil.

A repetition of these experiments in a furnace, on a larger scale, would undoubtedly have yet more successful results.

It deserves to be noticed that the curved form which these fibres of copper generally have is entirely favourable to the foregoing theory of their formation, and equally contrary to the supposition of their being produced by crystallization.

The power to which has been ascribed the phenomenon which forms the subject of these pages has hitherto been overlooked. It has not been considered what the effects might be of the contraction of a melted mass at the moment of its congelation. It is, however, a means of effects which may have acted on many occasions in the earth. Two matters of unequal fusibility, and of no attraction to each other, are not unlikely to have occurred blended in a state of fusion; and then the most fusible to have become pressed out from between the particles of the other when it solidified. If some evolved vapour had opened cavities in the mass, or rents had formed in it, the fluid matter will have escaped from the pressure into these voids, as has happened with the copper. If these receptacles for it have been wanting, it must have flowed to the external surfaces, and may have formed a crust there. The matter which lines or fills the cavities of some lavas has, perhaps, been so introduced into them.   •

A knowledge of the productions of art, and of its operations, is indispensable to the geologist. Bold is the man who undertakes to assign effects to agents with which he has no acquaintance ; which he never has beheld in action ; to whose indisputable results he is an utter stranger; who engages in the fabrication of a world alike unskilled in the forces and the materials which he employs.

AN ACCOUNT OF A NATIVE COMBINATION OF SULPHATE OF BARIUM AND FLUORIDE OF CALCIUM.

From Thomson's Annals of Philosophy, Vol. XVI, 1820, p. 48.

PARIS, *March* 24, 1820.

SIR : I acquired this substance in Derbyshire. It is many years since I ascertained its constitution. I have examined several minerals which in appearance bore a resemblance to it, but have not found any of them to be of the same nature. This species would hence appear to be of rare occurrence in the earth.

This substance formed a vein about an inch wide in a coarse shell limestone. Next to this substance was a layer of crystals of sulphuret of lead; and between these and the limestone rock a layer of crystals of carbonate of calcium.

I infer that these matters filled a vertical fissure in the limestone stratum ; and from the ideas I entertain of the mode by which such fissures have generally become occupied by their contents, I believe them to have been successively deposited in it by sublimation, either through the intense vehemence of subterranean fire, or by the agency of the vapour of water, or of some other gas.

This compound matter bears in its general appearance so strong a resemblance to fine compact grey limestone that the eye can probably not distinguish between them.

Forty-two grains of it lost 11.2 grs. in rain water at the

temperature of 61° Fahr.; consequently its density is 3.750. These 42 grs. of this stone by laying in the water did not absorb intó their substance a quantity of it equal to one-tenth of a grain.

It does not mark glass, and is readily scraped to a powder by a knife. It marked sulphate of barium. Its hardness and that of fluoride of calcium appeared to be the same.

It showed no electricity by heat. By friction it readily became electrified.

In the fire it lost no weight.

At the blow-pipe, it readily melted. The little bead while in fusion was transparent. On evolving, it became opaque. The transparency of the bead in a melted state is best seen with a very minute one. On fusing this matter long, it spreads on the coal, and becomes a refractory mass.

With borax, it dissolved with great effervescence into a brown glass. If much stone was used, the glass appeared quite black, but drawn out to a thread with the tongs, it was found to be of a fine hyacinth colour. These colours depend on the formation of sulphur.

With microcosmic salt it fused with effervescence to a clear colourless glass, which became opaque, and white on adding more of it.

A particle of this stone which had been *fused* on the charcoal being laid in a drop of water on a plate of silver, immediately made a black spot of sulphuret of silver on it.

This bit of melted stone, transferred to a drop of marine acid, on a piece of glass, partially dissolved with efferves-cence. The solution let exhale spontaneously, afforded crystals of chloride of barium.

Some of this stone in fine powder, being heated in a drop of sulphuric acid on a bit of glass, the polish of the glass was destroyed.

Water in which this stone in fine powder had been boiled was not affected by solution of nitrate of lead.

A bit of this stone, being heated in dilute marine acid, emitted a few bubbles of carbonic acid, but was not otherwise affected : 5.4 grs. of this mineral in very fine powder were let remain in an excess of marine acid till all action on them had ceased. The undissolved portion washed and gently ignited weighed 5.15 grs. The acid had acquired lime; so that this mineral contains a mechanical admixture of $\frac{4.6}{100.0}$ of carbonate of calcium.

This fine powder, which had been treated with the marine acid, had sulphuric acid evaporated to dryness on it in a platinum crucible. It was then digested in dilute marine acid. On evaporating this solution, a large quantity of sulphate of calcium in crystals was obtained.

From these results, sulphuric acid, fluorine, barytes, and lime, appear to be the elements of this mineral. It is consequently inferable that its proximate principles are sulphate of barium and fluoride of calcium.

The following experiments were made to obtain some idea of the proportions in which these two compound components of this mineral exist in it:

5.6 grs. of this stone in powder were heated in a platinum crucible in so large a quantity of sulphuric acid as to be entirely dissolved. The mixture was then exhaled dry, and ignited. The weight was now 7.85 grs. The increase had, therefore, been as $\frac{4.0}{100}$.

This augmentation of weight could arise only from the change of the fluoride of calcium into sulphate of calcium.

To know to what quantity of fluoride of calcium it corresponded, two grs. of pure fluoride of calcium in subtile powder were treated with sulphuric acid till the augmentation of weight ceased. The two grains had then become 3.65 grs.; accordingly the augmentation of weight was $= 1.65 = \frac{8.2}{100}$.

This Derbyshire mineral, therefore, consists of

Sulphate of barium	-	-	51.5
Fluoride of calcium	-	-	48.5
			100.0

Some error is created by the admixed carbonate of lime ; and which had not been removed.

This mineral presents us with a remarkable case of combination; that of a neutral salt with a body which is not a salt, but belongs to an order which is analogous to metallic oxides. I have met with another instance of the same kind. I have examined transparent crystals which were composed of anhydrous sulphate of calcium and chloride of sodium.

These combinations of their compounds may, however, perhaps, appear to some persons to cast doubts on the opinion that chlorine and fluorine are not acids.

These compounds will still be deserving of particular attention from consisting of *four* matters.

ON SOME CAPILLARY METALLIC TIN.

From Thomson's Annals of Philosophy, Vol. XVII—New Series, Vol. I—1821, p. 271.

PARIS, *February* 17, 1821.

SIR : M. Ampère, a few days ago, accidentally in conversation, mentioned a fact to me which much excited my attention, as it appeared to me completely to confirm the explanation I had ventured to offer of the mode of formation of the capillary copper in the slag of the Hartz, printed in the *Annals of Philosophy* for July, 1820.

For some purpose of the arts, Mr. Clement formed a cylinder of copper, and, to give it strength, introduced into it a hollow cylinder, or tube, of cast-iron. To complete the union of these two cylinders some melted tin was run between them. With the exact particulars of this construction, I am not acquainted, but the material circumstance is, that during the cooling of this heated mass, a portion of the melted tin was forced by the alteration of volume of the cylinders *through the substance* of the cast-iron cylinder, and issued over its internal surface in the state of *fibres*, which were curled and twisted in various directions. This form in the fibres of copper I had considered as very favourable to my hypothesis. Such was the tenuity of these fibres of tin that little tufts of them applied to the flame of a candle took fire, and burned like cotton.

This passage of melted tin through cast-iron has a perfect agreement with the passage of water by pressure through gold, and tends to elucidate and confirm the account of the celebrated Florentine experiment. Had the water on that occasion issued solid, it would have been in fibres.

This penetration of solid matters by fluids, by means of great mechanical force, will, perhaps, come to be thought deserving of more attention than has been yet paid to it; besides any scientific results to which the consideration of it may lead, it may be found to afford compound substances, not otherwise obtainable, and of value to the arts.

I am, sir, your most obedient servant,

JAMES SMITHSON.

ON THE DETECTION OF VERY MINUTE QUANTITIES OF ARSENIC AND MERCURY.

From Thomson's Annals of Philosophy, Vol. XX; New Series, Vol. IV, 1822, p. 127.

SIR: To be able to discover exceedingly small quantities of arsenic and mercury must, on many occasions, prove conducive to the purposes of the chemist and the mineralogist, more especially now that a very diminished scale of experiment, highly to the advantage of these sciences, is becoming daily more generally adopted.

But the occasion above all others in which the power of

doing this is important, are those of poisonings. In these it is often of the first moment to be able to pronounce with certainty, from portions of matter of extreme minuteness, on the existence and the nature of the poison.

Of Arsenic.

I have already communicated the method here proposed for the discovery of arsenic by employing it in the analysis of the compound sulphuret of lead and arsenic from Upper Valais, printed in the *Annals of Philosophy* for August, 1819, but not having mentioned the generality of its application, or the great accuracy of it, it seems not superfluous, from the importance of the subject, to resume it.

If arsenic, or any of its compounds, is fused with nitrate of potash, arseniate of potash is produced, of which the solution affords a brick-red precipitate with nitrate of silver.

In cases where any sensible portion of the potash of the nitre has become set free, it must be saturated with acetous acid, and the saline mixture dried and redissolved in water.

So small is the quantity of arsenic required for this mode of trial, that a drop of a solution of oxide of arsenic in water, which, at a heat of 54.5° Fahr. contains not above 1-80th of oxide of arsenic,* put to nitrate of potash in the platina spoon and fused, affords a considerable quantity of arseniate of silver. Hence when no solid particle of oxide of arsenic can be obtained, the presence of it may be established by infusing in water the matters which contain it.

The degree in which this test is sensible is readily determined.

With 5.2 grains of silver, I obtained 6.4 grains of arseniate of silver; but 0.65 grain of silver was recovered from the liquors, so that the arseniate had been furnished by 4.55 grs. of silver.

In a second trial 7.7 grains of silver, but of which only 6.8 grains precipitated, yielded 9.5 grs. of arseniate.

---

* Chimie de Thenard, ii, p. 167.

The mean is 140.17 from 100 of silver.

If we suppose 100 of silver to form 107.5 of oxide, we shall have

Oxide of silver	- -	107.50
Acid of arsenic	- -	32.67

Consequently 1 of acid of arsenic will produce 4.29 of arseniate of silver; 1 of white oxide of arsenic, 4.97; and 1 of arsenic, 6.56.

Of Mercury.

All the oxides and saline compounds of mercury laid in a drop of marine acid on gold with a bit of tin, quickly amalgamate the gold.

A particle of corrosive sublimate, or a drop of a solution of it, may be thus tried. The addition of marine acid is not required in this case.

Quantities of mercury may be rendered evident in this way which could not be so by any other means.

This method will exhibit the mercury in cinnabar. It must be previously boiled with sulphuric acid in the platina spoon to convert it into sulphate.

Cinnabar heated in solution of potash on gold amalgamates it.

A most minute quantity of metallic mercury may be discovered in a powder by placing it in nitric acid on gold, drying, and adding muriatic acid and tin.

A trial I made to discover mercury in common salt by the present method was not successful, owing, perhaps, to the smallness of the quantity, which I employed.

I am, sir, yours, &c.,

JAMES SMITHSON.

SOME IMPROVEMENTS OF LAMPS.

From Thomson's Annals of Philosophy, Vol. XX; New Series, Vol. IV, 1822, p. 363.

SIR : It is, I think, to be regretted, that those who cultivate science frequently withhold improvements in their apparatus and processes, from which they themselves derive advantage, owing to their not deeming them of sufficient magnitude for publication.

When the sole view is to further a pursuit of whose importance to mankind a conviction exists, all that can do so should be imparted, however small may appear the merit which attaches to it.

Of the Wicks of Lamps.—The great length of wick commonly put to lamps for the purpose of supplying the part which combustion destroys, is, on several accounts, extremely inconvenient. It occupies much space in the vessel, and requires an enlargement of its capacity; it is frequently the occasion of much dirt, &c. This great length of wick is totally unnecessary.

Fig. 1.   Fig. 2.   Fig. 3.   Fig. 4.

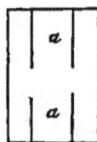

It is advantageously supplied by a tube containing a bit of cotton wick about its own length, or some cotton wool, fig. 1, and at the end of which is placed a stout bit of wick or cotton wool, fig. 2.

This loose end receives a supply of oil from the cotton under it with which it is put into contact, and when it becomes burned, it is easily renewed.

A loose ring of wick may in like manner be applied to the argand lamp. This removes the necessity of the long tube into which the wicks, now used, descend, and thus greatly contracts this lamp in height.

Of Wax Lamps.—Oil is a disagreeable combustible for small experimental purposes, and more especially when lamps are to be carried in travelling. I have, therefore, substituted wax for it. I experienced, however, at first, some difficulty in accomplishing my object.

The wicks of my lamps are a single cotton thread, waxed by drawing through melted wax. This wick is placed in a burner made of a bit of tinned iron sheet, cut like fig. 3, and the two parts *a a* raised into fig. 4.

This burner is placed in a china cup, about 1.65 inches in diameter, and 0.6 in. deep. Fragments of wax are pressed into this cup. But great care must be taken that each time the lamp is lighted, bits of wax are heaped up in contact with the wick, so that the flame shall immediately obtain a supply of melted wax. This is the great secret on which the burning of wax lamps depends.

When the wick is consumed, the wax must be pierced with a large pin down to the burner, and a fresh bit of waxed cotton introduced.

I employ a wax lamp for the blowpipe. This has, of course, a much larger wick, and this wick has a detached end to it, as above described.

Extinguishing Lamps.—The best way of doing this is to extinguish the ignited part of the wick by putting sound wax on to it, and then blowing the flame out. This preserves the wick entire for future lighting again.

This mode applied to candles is much preferable to the use of an extinguisher, or douters, to which there are many objections.

ON THE CRYSTALLINE FORM OF ICE.

From Thomson's Annals of Philosophy, Vol. XXI; New Series, Vol. V, 1823, page 340.

March 4, 1823.

SIR : I have just seen a memoir in the Annales de Chimie et de Physique for Oct. 1822, but published about a month ago, on the crystalline form of ice.

Mr. Hericart de Thury is said to have observed ice in hexagonal and triangular prisms ; and Dr. Clarke, of Cambridge, in rhomboides of 120° and 60°.

M. Haüy supposed the form to be octahedral, and so did Romé de l'Isle ; and, if I mistake not much, there is in an ancient volume of the Journal de Physique by Rozier, an account of ice in acute octahedrals.

Are these accounts and opinions accurate ?

Hail is always crystals of ice more or less regular. When they are sufficiently so to allow their form to be ascertained, and which is generally the case, it is constantly, as far as I have observed, that of two hexagonal pyramids joined base to base, similar to that of the crystals of oxide of silicium or quartz, and of sulphate of potassium. *One of the pyramids is truncated*, which leads to the idea that ice becomes electrified on a variation of its temperature, like tourmaline, silicate of zinc, &c.

I do not think that I have measured the inclination of the faces more than once. The two pyramids appeared to form by their junction an angle of about 80 degrees.

Snow presents in fact the same form as hail, but imperfect. Its flakes are skeletons of the crystals, having the greatest analogy to certain crystals of alum, white sulphuret of iron, &c., whose faces are wanting, and which consist of edges only.

In spring and autumn ; that is, between the season of

snow and that of hail, the hail which falls partakes of the nature of both, is partly the one and the other; its crystals, though regular, are opaque, of little solidity, and consist, like snow, of an imperfect union of grains, or smaller crystals.

A MEANS OF DISCRIMINATION BETWEEN THE SULPHATES OF BARIUM AND STRONTIUM.

From Thomson's Annals of Philosophy, Vol. XXI; New Series, Vol. V, 1823, page 359.

April 2, 1823.

SIR: To distinguish barytes and strontian from one another, it is directed in No. 19 of the Journal of the Royal Institution to dissolve in an acid which forms a soluble salt with them, to decompose by sulphate of soda, and to add subcarbonate of potash to the filtered liquor. If the earth tried is strontian, a precipitate falls; if barytes, not.

When these matters are in a state to be soluble in an acid, a more certain, I apprehend, and undoubtedly a much easier proceeding, is to put a particle into a drop of marine acid on a plate of glass, and to let this solution crystallize spontaneously. The crystals of chloride of barium in rectangular eight-sided plates are immediately distinguishable from the fibrous crystals of chloride of strontium.

I have not repeated the process above quoted; but if sulphate of strontium did possess the solubility in water there implied, this quality presented a ready method by which mineralogists would be enabled to distinguish it from sulphate of barium. On trial I did not find water, or solution of sulphate of soda, in which sulphate of strontian had long lain, produce the least cloud on the addition of what is called subcarbonate of soda.

6

The means I have long employed to distinguish the two sulphates apart was to fuse with carbonate of soda, wash, dissolve in marine acid, &c.; but this process requires more time and trouble than is always willingly bestowed, and may even present difficulties to a person not familiarized with manipulations on very small quantities.

A few months ago a method occurred to me divested of these objections. The mineral in fine powder is blended with chloride of barium, and the mixture fused. The mass is put into spirit of wine, whose flame is coloured red if the mineral was sulphate of strontium. The red colour of the flame is more apparent when the spirit is made to boil while burning, by holding the platina spoon containing it over the lamp.

ON THE DISCOVERY OF ACIDS IN MINERAL SUBSTANCES.

From Thomson's Annals of Philosophy, Vol. XXI; New Series, Vol. V, 1823, page 384.

April 12, 1823.

SIR : Acids, it is well known, have been repeatedly overlooked in mineral substances, and hence dubiousness still hovers over the constitution of many, although they have formed the subjects of analysis to some of the greatest modern chemists.

To be able to dissipate all doubts—to ascertain with certainty whether an acid does or does not exist, and, if one is present, its species, and this with such facility that the trial may be indefinitely renewed at pleasure, and made by all, so that none need believe but on the testimony of his own experiments, is the degree of analytical power which it would be desirable to possess.

So far as I have gone in these respects, I here impart.

As the carbonates of soda and of potash precipitate all the solutions of earths and metals in acids, so do they decompose all their salts by fusion with them. Fusion with carbonate of soda or potash affords there a general method of separating acids from all other matters.

Lead forms an insoluble compound with all the mineral acids except the nitric. It may consequently be immediately known whether a mineral does or does not contain an acid element by the carbonate of soda or potash, with which it has been fused after saturation by acetous acid, forming or not forming a precipitate with a solution of lead.

If the production of a precipitate proves the presence of an acid, the determination of its species will present no great difficulty.

1. *Sulphuric Acid.*—If the alkali which has received it from the mineral is fused on charcoal, and then laid in a drop of water placed on silver, a spot of sulphuret of silver will be produced, as I have stated on a former occasion.* Bright copper will likewise serve for this purpose.

Fusion in the blue flame will often be sufficient to deoxidate the sulphur.

It is needless to observe that the alkali used in this trial must itself be perfectly free from sulphuric acid. When such is not possessed, its place may be supplied by Rochelle salt, or by cream of tartar.

2. *Muriatic Acid.*—I have likewise discovered a test of chlorine, and consequently of muriatic acid, of delicacy equal to the foregoing. If any matter containing chlorine or muriatic acid is laid on silver in a drop of solution of yellow sulphate of iron, or of common sulphate of copper, a spot of a black chloride of silver, whose colour is independent of light, and which has not been attended to ·by chemists, is produced. The chlorine in a tear, in saliva, even in milk, may be thus made evident. When the quantity of chlorine in a liquor is very small, a bit of sulphate

---

* Annals of Philosophy for July, 1820.

of copper placed in it on the silver is preferable to a solution. To find chlorine in milk, I put some sulphate of copper to it, and placed a small piece of bright silver in the mixture.

3. *Phosphoric Acid.*—The alkali containing it, after saturation by acetous acid, gives a sulphur-yellow precipitate with nitrate of silver, which no other acid does. The precipitate obtained with lead crystallizes on the blow-pipe. M. Berzelius's elegant method of detecting phosphoric acid is universally known.

4. *Boracic Acid.*—Its presence in carbonate of magnesia, and in some other of its compounds, is indicated by the green colour they give, during their fusion, to the flame of the lamp.

M. Gay-Lussac has observed that a solution of boracic acid in an acid changes the colour of turmeric paper to red, like an alkali.* Borax, to which sulphuric acid has been put, does so, and the same is of course the case with a bead of soda containing boracic acid.

The most certain test of boracic acid in a soda bead, &c., is to add sulphuric acid to it and then spirit of wine, whose flame is coloured green, if boracic acid is present.

5. *Arsenical Acid.*—Alkali containing it produces a brick-red precipitate with nitrate of silver.†

6. *Chromic Acid.*—Chromate of soda and its solution are yellow, and so is the precipitate with lead. That with silver is red.

Chromate of soda or potash fused on a plate of clay leaves green oxide of chromium.

Chromate of lead fused on a plate of clay produces a very dark-green mass, which is probably chromate of lead; with an addition of lead, it forms a fine red, or orange glass.

Lead added to the green oxide left by chromate of soda

---

* Annales de Chimie et de Physique, tomo xvi. p. 76.
† *Annals of Philosophy*, N. S. vol. iv. p. 127.

on the clay plate, dissolves it, and forms an orange-coloured glass.

The green oxide of chromium sometimes acts the part of an acid. I have seen a combination of it with oxide of lead found in Siberia, in regular hexagonal prisms, having the six edges of the terminal face truncated (Haüy, pl. lxviii. fig. 63); melted with lead on the clay plate this would undoubtedly produce the orange glass; and fused with nitrate of potash it would form chromate of potash.

7. *Molybdic Acid.*—If molybdate of soda or potash, or, I apprehend, any other molybdate, is heated in a drop of sulphuric acid, the mixture becomes of a most beautiful blue colour, either immediately, or on cooling.

The solution of molybdate of soda in sulphuric acid affords with martial prussiate of potash, a precipitate of the same colour that copper does. Tincture of galls gives with this acid solution a green precipitate; but with an alkaline solution of molybdic acid galls produce a fine orange precipitate. If an alkali is put to the green precipitate, it becomes orange; and if an acid to the orange precipitate, it becomes green.

8. *Tungstic Acid.*—If tungstate of soda is heated with sulphuric acid, the granules of precipitated tungstic acid become blue, but not the solution; and the phenomena cannot be confounded with those presented by molybdate of soda. Martial prussiate of potash has no effect on this acid liquor.

Tincture of galls put to the solution of tungstate of soda in water does not affect it. On the addition of an acid to this mixture, a brown precipitate forms.

If tungstate of soda is heated to dryness with a drop of muriatic acid, a yellow mass is left. On extracting the saline matter by water, yellow acid of tungsten remains. It is readily soluble in carbonate of soda. If taken wet on the blade of a knife, it soon becomes blue. This is made very evident by wiping the blade of the knife with a bit of white paper. Possibly a small remainder of muriatic or sulphuric acid among it is required for this effect.

9. *Nitric Acid.*—Nitrate of ammonia produces no deflagration when filtering paper, wetted with a solution of it and dried, is burned; the salt volatilizing before ignition, most, or all, the other nitrates deflagrate.

If metallic copper is put into the solution of a nitrate, sulphuric acid added, and heat applied, the copper dissolves with effervescence.

10. *Carbonic Acid.*—It is to be discovered in the mineral itself. The application of heat is, in some cases, required to render the effervescence sensible. It has been sometimes overlooked in bodies from want of attention to this circumstance.

11. *Silica.*—A simple and sufficient test of it is the formation of a jelly, when its combination with soda is put into an acid.

It has evidently not been intended to enumerate all the means by which the presence of each acid in the soda bead could be perceived or established. Little has been said beyond what appeared required and sufficient.

Mention has been made above of small plates of clay.

They are formed by extending a white refractory clay by blows with the hammer, between the fold of a piece of paper, like gold between skins. The clay and paper, are then cut together with scissars into pieces about 4-10ths of an inch long, and 2½-10ths of an inch wide, and hardened in the fire in a tobacco-pipe.

They are very useful additions to the blowpipe apparatus. They admit the use of a new test, oxide of lead. They show to great advantage the colours of matters melted with borax, &c. Quantities of matter too minute to be tried on the coal, or on the platina foil, or wire, may be examined on them alone, or with fluxes. Copper may be instantly found in gold or silver by fusing the slightest scrapings of them with a little lead, &c., &c.

Cut into very small, very acute triangles, clay affords a substitute for Saussure's sappare.

AN IMPROVED METHOD OF MAKING COFFEE.

From Thomson's Annals of Philosophy, Vol. XXII; New Series, Vol. VI, 1823, page 30.

June 4, 1823.

SIR: From the highly fugacious nature of that part of coffee on which its fine flavour depends, a practice has become very generally adopted of late years of preparing the liquor by mere percolation.

This method has not only the great defect of being excessively wasteful, but the coffee is likewise apt to be cold.

Coction and the preservation of the fragrant matter are, however, not inconsistent. The union of these advantages is attainable by performing the operation in a close vessel. To obviate the production of vapour, by which the vessel would be ruptured, the boiling temperature must be obtained in a water-bath.

In my experiments I made use of a glass phial closed with a cork, at first left loose to allow the exit of the air. Cold water was put to the coffee.

This process is equally applicable to tea.

Perhaps it may also be employed advantageously in the boiling of hops, during which, I understand, that a material portion of their aroma is dissipated; as likewise possibly for making certain medical decoctions.

This way of preparing coffee and tea presents various advantages. It is productive of a very considerable economy, since by allowing of any continuance of the coction without the least injury to the goodness, all the soluble matter may be extracted, and consequently a proportionate less quantity of them becomes required. By allowing the coffee to cool in the closed vessel, it may be filtered through paper, then returned into the closed vessel, and heated again, and thus had of the most perfect clearness without any foreign addition to it, by which coffee is impaired. The liquors may be kept for any length of time at a boiling heat, in private families, coffee houses, &c., so as to be ready at the very instant called for.

It will likewise prove of no small conveniency to travellers who have neither kettle, nor coffee-pot, nor tea-pot, in places where these articles are not to be procured, as a bottle will supply them.

In all cases means of economy tend to augment and diffuse comforts and happiness. They bring within the reach of the many what wasteful proceedings confine to the few. By diminishing expenditure on one article, they allow of some other enjoyment which was before unattainable. A reduction on quantity permits indulgence in superior quality. In the present instance, the importance of economy is particularly great, since it is applied to matters of high price, which constitute one of the daily meals of a large portion of the population of the earth.

That in cookery also, the power of subjecting for an indefinite duration to a boiling heat, without the slightest dependition of volatile matter, will admit of beneficial application, is unquestionable.

A DISCOVERY OF CHLORIDE OF POTASSIUM IN THE EARTH.

From Thomson's Annals of Philosophy, Vol. XXII; New Series, Vol. VI, 1823, page 258.

SIR: A RED ferruginous mass, containing veins of a white crystalline matter, part of a block which was said to have been thrown out of Vesuvius during a late eruption, was brought to me, with a request that I would tell what it was.

This red ferruginous rock was a spongy lava, in the substance of which was here and there lodged a crystal of augite or pyroxene of Haüy, or of hornblende.

The white matter filled most of the larger cavities, and was more or less disseminated through nearly the whole of the mass.

It had a saline appearance; a tabular fracture could be seen in it with a lens, and in some few places regular cubical crystals were discernible.

I supposed it to be chloride of sodium, or muriate of ammonia.

Heated in a matrass, it decrepitated slightly, and melted, but little or nothing sublimed.

This white matter dissolved entirely in water. Laid on silver with sulphate of copper, it produced an intense black stain.

Chloride of barium added to the solution caused only a very slight turbidness, due probably to some sulphate of lime which is present.

Tartaric acid occasioned an abundant formation of crystals of tartar. Chloride of platinum immediately threw down a precipitate, and distinct octahedral crystals of the same nature afterwards appeared.

On decomposition by nitric acid, only prismatic crystals of nitrate of potash could be perceived. On a second crystallization, a few rhombic crystals were discovered; but nitrate of potash sometimes presents this form.

It appears from these experiments, that this white saline matter is pure, or nearly pure, chloride of potassium.

I am inclined to attribute its introduction into the lava to sublimation.

As chloride of potassium is a new species in mineralogy, I shall send the specimen to the British Museum.

A METHOD OF FIXING PARTICLES ON THE SAPPARE.

From Thomson's Annals of Philosophy, Vol. XXII; New Series, Vol. VI, 1823, page 412.

October 24, 1823.

SIR: When the species of minerals are ascertained by their physical qualities, they mostly undergo no injury, or but a very slight one; as that attending the determination of their hardness, the colour of their powder, their taste, &c. This is certainly a material advantage, and would highly recommend this method, was it constantly adequate to its purpose. That it is not so, however, we have a proof in the great errors into which have fallen those best skilled in it. Mr. Werner, its principal and most distinguished professor, was unable by its means to discover the identity of the jargon and the hyacinth; of the corundum and the sapphire; of his apatite and his spargelstein; and while he thus parted beings, as it were, from themselves, he forced others together which had nothing in common.

The chemical method justly boasts its certainty; but it carries destruction with it, and often bestows the knowledge of an object only at the expense of its existence. The sole remedy which can be opposed to this defect is to reduce the scale of operating; and thus render the sacrifice which must be made as small as it is possible.

M. de Saussure's* ingenious contrivance for subjecting the most minute portions of matters to fire, by fixing them on a splinter of sappare, appeared to fulfil the conditions of this problem, and to have accomplished all that could be desired. It has, however, been scarcely at all employed, owing to the excessive difficulty in general of making the particles adhere; and in consequence the almost unpossessed degree of patience required for, and time consumed by, nearly interminable failures.

That such should be the case could not but be a subject of much regret, for besides economy of matter, of time, of labour, and the great beauty of deriving knowledge from so diminutive a source, and attaining important results with such feeble agents; reduction of volume became, in this instance, productive of increase of power, and thence, of an extension of the series of qualities by which substances are characterised.

A slight alteration which I have made in M. de Saussure's process has removed the objection to it. To water, saliva, gum water, which he employed, the last of which is not sensibly superior to the former, I have substituted a mixture of water and refractory clay.

Small triangles, or slender strips, of baked clay may be used in lieu of sappare, which is not at all times to be procured; or a little of the moist clay may be taken up on the end of a platina, or other wire, and the object to be tried touched with it. This way may be applied to pieces of the ordinary size, and supersede the use of the platina tongs.

But a proceeding which I have only recently adopted appears to deserve the preference. Almost the least quantity of clay and water is put on the *very end* of a platina wire, filed flat there. With this, the particle of mineral lying on the table can be touched in any part chosen; for a moment

---

* Journal de Physique, par Rozier, tome 45.

or two it is dry, and may be taken up, and put into the flame, without the clay exploding, as not unoften happens when more of it is used. Particles of the least visible minuteness may be thus submitted to trial with the utmost facility. The contact of the particle with the wire may, in general, be so managed as to be extremely slight, as the slenderest point is sufficient to support it. However, when the utmost heat possible is desired, a fragment of a less conducting matter may, if deemed necessary, be interposed.

There may be cases in which the presence of .the clay is objectionable. I conceived that some of the body itself to be tried, would on these occasions, supply its place. Flint was the least promising of any in this respect. It was selected for the experiment. With a paste of its powder and water, pieces of flint were successfully cemented to flint, and some of this paste taken on the end of a wire, served, if not quite as well as clay, yet very sufficiently. After several times igniting and quenching in cold water, the reduction of very hard matters to subtile powder is attended with no difficulty.

Earth of alum would perhaps be preferable to pipe-clay for making the triangles on strips, and for agglutinating objects to them. It would even have the advantage over sappare of being a simple substance. Some from the Paris shops acquired only little solidity in the fire; but I afterwards learned that it had been obtained from alum by fire.

Since I have been in possession of this means of so effectually confining the subjects of examination as to be able to continue during pleasure to act on them, I have directed but little attention to the fusibility of matters. Quartz, whose fusion has been called in question by M. Berzelius,* has seemed to be quite refractory. On some few occasions when it has proved otherwise, the phenomena have neither corresponded with M. de Saussure's account, nor been always the same, which certainly admits of the fusion being attributed to an accidental cause.

---

* De l'emploi du Chalumeau, p. 108.

But I have found with much surprise that flint can be melted without difficulty; and even of a considerable bulk. Where the heat is most intense, a degree of frothing takes place; where it is less, there is a swelling of parts of the surface. The effects were the same with French and English flint, with black and with horn-coloured. Does flint, like pitchstone, contain bitumen, which, at a certain heat, tends to tumefy it? This might explain the smell from its collision, and the oil which Neumann obtained by its distillation, and to which no credit has been ever given. No doubt can, I conceive, be entertained of flint being a volcanic production. On this point I may speak again at a future opportunity.

In using mere water, diamond, anthracite, plumbago, were particularly difficult of trial, as any adhesion they had contracted with the sappare was quickly destroyed by the combustion of their surface, while, as the intention in their case is not to subject to great heat, they may be so secured in the clay as at least very much to retard their escape. Here acting on very minute particles is essential, as when large pieces are employed, the effect is too slow to be perceptible.

A pleasing way of demonstrating the combustion of plumbago, and of even exhibiting the iron in it, is to rub a little from the wetted point of a pencil on one of the clay plates mentioned in a former paper.*

In trying diamond it was imagined that its glow continued an unusual time after removal from the fire. The present method afforded the means of making a comparison. A fragment of diamond, and another of quartz, chosen purposely of rather a larger size, were fixed near each other in the clay; and it was observed that the diamond was most luminous while under the action of the flame, and longer so after removal from it. Its being a very slow conductor of heat may occasion in part the latter quality.

* *Annals* for May.

In the same way the unequal fusibility of two substances
may probably, on some occasions, be ascertained; and serve
from deficiency of a better, as a means of distinction between them.

I am, sir, yours, &c.

J. SMITHSON.

ON SOME COMPOUNDS OF FLUORINE.

From Thomson's Annals of Philosophy, Vol. XXIII; New Series, Vol.
VII, 1824, p. 100.

January 2, 1824.

SIR: When numberless persons are seen, in every direction, pursuing a subject with the utmost ardour, it is natural to conclude that their labors have accomplished all that
was within their reach to perform.

It must, therefore, in mineralogy be supposed, that those
substances whose abundance has placed them in every hand,
have been fully scrutinized, and are thoroughly understood;
and that if now to extend the boundaries of the science it
is not indispensable to explore new regions of the earth,
and procure matters hitherto unpossessed, it is yet only to
objects the most rare, the most difficult of acquisition, that
inquiry can be applied with any hope of new results.

A want of due conviction that the materials of the globe
and the products of the laboratory are the same, that what
nature affords spontaneously to men, and what the art of
the chemist prepares, differ no ways but in the sources from
whence they are derived, has given to the industry of the
collector of mineral bodies an erroneous direction.

What is essential to a knowledge of chemical beings has
been left in neglect; accidents of small import, often of
none, have fixed attention—have engrossed it; and a fertile

field of discovery has thus remained where otherwise it would have been exhausted.

Fluor spar has decorated mineral cabinets from probably the earliest period of their existence; every tint with which chance can paint it; each casual diversity of form and appearance under which it may present itself have been long familiar, and its true nature continues a problem; and its decomposition by fire was yet to be learned.

Fluor Spar.

If a very minute fragment of fluor spar is fastened by means of clay* to the end of a platina wire nearly as fine as a hair, which is the size I now employ even with fluxes, it will be perceived on the first contact of the fire to melt with great facility. As the fusion is prolonged, the fusibility will decrease; protuberances will rise over the surface of the ball; it will put on what is designated by the term of the cauliflower form; and finally become entirely refractory. On detaching it from the wire, it will prove hollow. This little capsula being taken up again by its side, and its edge presented to the flame, thin and porous as this edge is, it will withstand its utmost violence.

Such an alteration of qualities proclaims an equal one of nature. I had no doubt that the calcium had absorbed oxygen, and parted with fluorine; that the mass had ceased to be fluor spar, and was become quicklime. On placing it in a drop of water my conjecture was confirmed; a solution took place by which test papers were altered; a *cremor calcis* soon appeared; and on allowing the mixture to become spontaneously dry, a white powder remained, which acids dissolved with effervescence.

That the fluoric element was gone admitted not of doubt. To pursue it in its escape; to coerce it, and render it palpable to the senses, could not be required to establish the fact. It may, however, be done.

---

* *Annals* for December.

The open tube described by M. Berzelius in his valuable
work on the blowpipe, is adapted to the purpose by an addition to it. A small plate of platina foil, or a curved plate
of baked clay, is introduced a little way into one of its ends;
and secured by bringing with the point of the flame the glass

into contact with it.   The body to be tried is fixed to this
plate by means of moist clay; and may then be subjected
for any time to any degree of heat.

Thus tried, fluor spar quickly obscured the glass by a thick
crust of siliceous matter; and coloured yellow a bit of paper
tinged with logwood.

M. Berzelius assigns fernambuc wood for the test of fluoric acid.   Bergman says that this wood affords a red infusion which alkalies turn blue.*   None such could be
procured, but it was found that logwood might be substituted for it.   The paper tinged with this, like that mentioned
by M. Berzelius, is made yellow by fluoric acid and oxalic
acid; but it did not seem to be so by sulphuric or muriatic
acids, nor by phosphoric acid.

Topaz.

In extremely minute particles, topaz subjected to the fire
at the end of a very slender wire soon becomes opaque and
white; but I perceived no marks of fusion.

This change is undoubtedly occasioned by the loss of its
fluoric part.   One of the times I was at Berlin, M. Klaproth
gave me, as his reason for not publishing the analysis of
topaz, that in the porcelain furnace it sustained a great loss
of weight, the cause of which he had not then been able to
ascertain.

Topaz ground to impalpable powder, and blended with
carbonate of lime, melted with ease.   Some of this mixture

---

* Analysis of Mineral Waters.

fused on the platina plate at the mouth of the tube, made an abundant deposit of silica over its interior surface; and the bit of logwood paper at the end of it had its blue colour altered to yellow.

In the trial in this way of substances of difficult fusion, an apparatus of the following construction is more favourable than the one above described.

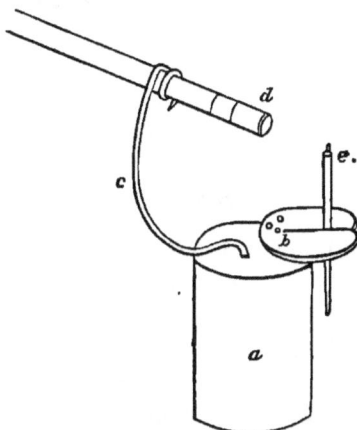

a. A bottle cork.

b. A slice of the same fixed with three pins.

c. A wire.

d. A cylinder of platina foil introduced into the mouth of the glass tube, to prevent its being softened and closed by the flame.

e. A platina wire, at the end of which is cemented with clay the subject of trial.

I formerly suggested that topaz might be a compound of silicate of alumina, and of fluate of alumina.* I am now convinced that no oxygen exists in it; but that it is a combination of the fluorides of silicium and aluminum.

This system produces a considerable alteration in the proportions of its elements.

7

* Philosophical Transactions for 1811.

The mean of the six analyses quoted by M. Haüy, in the second edition of his Mineralogy, is

Silica	-	-	-	36.0
Alumina	-	-	-	52.3
Fluoric acid		-	-	9.7
			98.0

Deducting the oxygen from the metals, we have

Silicium	-	-	-	18.0
Aluminium	-	-	-	27.7
Fluorine	-	-	-	52.3
			98.0

Kryolite.

It has been observed to diminish in fusibility during fusion,* and it was in every respect probable, from what had been seen with the foregoing bodies, that it would be decomposed in the fire. After being kept some time melted, it afforded an alkaline solution, which, by exposure to the air, became carbonate of soda, effloresced, effervesced with nitric acid, and produced crystals of nitrate of soda.

Fused on the platina plate at the mouth of the tube; a copious deposit of silex collected in the tube; and the bit of logwood paper became very yellow.

Kryolite heated in sulphuric acid on glass destroyed its polish.

---

1. These experiments render it highly probable that fluorine will be expelled from every compound of it by the agency of fire; and consequently that we are now in possession of a general method of discovering its presence in bodies. In cases where a matter is infusible, and parts with it with great difficulty, as in that of topaz, it may be required to reduce it to fine powder, or to act upon it by some ad-

---

* Haüy's Mineralogy.

mixture with which it melts, for the sake of promoting division and multiplying surfaces.

Hereby is supplied what may have seemed to be an omission in the paper on acids.* Although it was not such, since fluorine is not an acid; and fluoric acid may never occur in a mineral substance; as it can probably exist in combination only with ammonia; all its other supposed compounds being doubtless fluorides.

2. The theory of these decompositions may be acquired by experiment; and light obtained on the nature of the compounds.

If fluor spar, for instance, is a combination of oxide of calcium and fluoric acid, and this is expelled from the oxide merely by the force of fire, the decomposition of it will take place in closed vessels without the presence of oxygen or of water; fluoric acid will be obtained; and the weight of this acid and the lime will be equal together to that of the original spar.

If the spar is metallic calcium and fluorine, and when heated in oxygen absorbs this, and parts with fluorine, it is fluorine which will be collected in the vessels, and its weight and that of the lime will together exceed that of the spar by the oxygen of the lime.

If it is water which is the agent of decomposition, fluoric acid will be collected; but here the excess of weight will not only equal the oxygen absorbed by the lime, but also the hydrogen which has acidified the fluorine; and this increased weight of the fluoric acid will prove that hydrogen is an element of it.

It appears to have been fluoric acid which in the above related experiments passed into the tubes; but the inflammable matter of the flame would probably have rendered emitted fluorine such. It becomes of high importance to ascertain whether ignited fluor spar is decomposed by passing water over it, and if so what are the products. It is

---

* *Annals* for May.

not convenient to myself at present to make the experiment :
I therefore resign it to others.

How far the difficulty which the action of fluorine on the
vessels in which it is contained, as opposed to its examination, would be obviated by employing vessels of its compounds, as of fluor spar, or of chloride of silver; or whether
it acts on all oxides as it does on silica, experiments have
not informed me.

3. The vegetation of matters before the blow-pipe is
attributed by a great chemist to a "new state of equilibrium induced by heat between the constituent parts of
bodies,"* but the phenomena do not accord with the explanation.

Was such the cause of the acquired infusibility, it would
manifest itself through the whole mass as soon as fusion had
enabled the new arrangement. It is, on the contrary confined to the surface; the interior portion continues fluid;
but wherever any of this bursts the shell, and issues forth,
it is instantly fixed in immovable solidity; and when the
process has attained its final state, a hollow globule remains.

Why is the change of quality limited to the surface; how
has been produced the central cavity; what has forced away
the matter which occupied it? A new element has been
received from without, one which existed in the matter has
been parted with in a state of vapour. This double action
may probably be inferred wherever a matter presents this
species of vegetation.

Some metallic bodies, as tin, lead, sulphuretted tin, arsen-
icated nickel, &c., present another species of vegetation,
caused by the absorption of oxygen, and the production
over their surface of a matter more bulky than the metal
from which it is produced, and infusible at the heat to
which it is exposed. Here no internal void forms.

The mode of fusion of epidote had led me to suspect the
existence of fluorine in it; but on trial with the second ap-

---

* De l'Emploi du Chalumeau, p. 94.

paratus, represented above, I could not perceive a trace of it. A more accurate observation of its fusion has shown me that it does not, as generally supposed, form the cauliflower. It appears to do so only where so large a mass is exposed to the fire that but points of its surface are fused in succession. If a very minute bit is employed, it is clearly seen to puff up like borax, stilbite, &c.; and then, like them, become less fusible; from the separation, doubtless, of a vapourized element on which its greater fusibility had depended. The smallest particle of fluor spar shows no such inflation.

We see here three several cases of intumescence in the fire: one where a gas is absorbed; one where a gas, or vapour, is disengaged; one where the two effects are concomitant.

There may be persons who, measuring the importance of the subject by the magnitude of the objects, will cast a supercilious look on this discussion; but the particle and the planet are subject to the same laws; and what is learned upon the one will be known of the other.

AN EXAMINATION OF SOME EGYPTIAN COLOURS.

From Thomson's Annals of Philosophy, Vol. XXIII, New Series, Vol. VII, 1824, p. 115.

January 2, 1824.

SIR: More than commonly incurious must he be who would not find delight in stemming the stream of ages: returning to times long past, and beholding the then state of things and men.

In the arts of an ancient people much may be seen concerning them: the progress they had made in knowledge of various kinds; their habits; their ideas on many subjects.

And products of skill may likewise occur, either wholly unknown to us, or superior to those which now supply them.

I received from Mr. Curtin, who travelled in Egypt with Mr. Belzoni, a small fragment of the tomb of King Psammis. It was sculptured in basso relievo which were painted.

The colours were white, red, black and blue.

I have heard the white of Egyptian paintings extolled for its brilliancy and preservation. I found the present to be neither lead nor gypsum; but carbonate of lime. Chlorides of barium caused no turbidness in its solution. An entire sarcophagus of arragonite proves that the ancient Egyptians were in possession of an abundant store of this matter, remarkable often for its perfect whiteness. Was it the material of their white paint?

The red was oxide of iron. By heating, it became black, and returned on cooling to its original hue. In a case where so much foreign admixture was present, since the layer of red was much too thin to allow of its being isolated, I considered this as a better proof of red oxide of iron than obtaining prussian blue.

The black was pounded wood charcoal. After the carbonate of lime with which it was mixed had been removed by an acid, the texture of the larger particles were perfectly discernible with a strong lens; and in the fire it burned entirely away.

The blue is what most deserves attention. It was a smalt, or glass powder, so like our own, though a little paler, as to be mistaken for it by judges to whom I showed it; but its tinging matter was not cobalt, but copper. Melted with borax and tin, the red oxide of copper immediately appeared.

Many years ago I examined the blue glass with which was painted a small figure of Isis, brought to me from Egypt by a relation of mine, and found its colouring matter to be copper.

I am informed that a fine blue glass cannot at present be

obtained by means of copper.   What its advantages would be above that from cobalt, it is for artists to decide.

Intent upon the blue smalt, it unfortunately did not occur to me to examine, till I had washed nearly the whole of it away to waste, what was the glutinous matter which had been so true to its office for no less a period than 3,500 years; for the colours were as firm on the stone as they can ever have been.

A small quantity of it recovered from the water did not seem to form a jelly on concentrating its solution; or to produce a precipitate with galls.   I imagined its vegetable nature ascertained by its ashes restoring the colour of reddened turnsol paper, till I found those of glue do the same.

The employment of powder of charcoal for a black would seem to imply an unacquaintance with lamp-black, and, perhaps, with bone black, and that of copper to colour glass blue, a deficiency of cobalt.   And if the glutinous matter should prove, on a future examination, to be vegetable, our glue being then possessed may, perhaps, be deemed questionable.

SOME OBSERVATIONS ON MR. PENN'S THEORY CONCERNING THE FORMATION OF THE KIRKDALE CAVE.

From Thomson's Annals of Philosophy, Vol. XXIV; New Series, Vol. VIII, 1824, p. 50.

June 10, 1824.

SIR: No observer of the earth can doubt that it has undergone very considerable changes.   Its strata are everywhere broken and disordered; and in many of them are enclosed the remains of innumerable beings which once had life; and these beings appear to have been strangers to the climates in which their remains now exist.

In a book held by a large portion of mankind to have been written from divine inspiration, an universal deluge is recorded. It was natural for the believers in this deluge to refer to its action, all, or many, of the phenomena in question; and the more so as they seemed to find in them a corroboration of the event.

Accordingly, this is what was done, as soon as any desire to account for these appearances on the earth became felt. The success, however, was not such as to obtain the general assent of the learned; and the attempt fell into neglect and oblivion.

Able hands have lately undertaken the revival of this system; Mr. Penn has endeavoured to reconcile it with the facts of the Kirkdale Cave, which appeared to be strongly inimical to it.

Acquainted with Mr. Penn's opinions only from the "Analysis of the Supplement to the Comparative Estimate" in the Journal of the Royal Institution for January, not having seen this Supplement itself, the Comparative Estimate, nor even a review of this in a former number of the Journal, and knowing of Mr. Buckland's *Reliquiæ Diluvianæ*, only the account of the Kirkdale Cave published in the Philosophical Transactions for 1822, I have hesitated long about communicating the present observations, which presented themselves during the perusal of the above-mentioned slender abstract.

I have yielded to a sense of the importance of the subject in more than one respect, and of the uncertainty when I shall acquire ampler information at more voluminous sources—to a conviction that it is in his knowledge that man has found his greatness and his happiness, the high superiority which he holds over the other animals who inhabit the earth with him, and consequently that no ignorance is probably without loss to him, no error without evil, and that it is therefore preferable to urge unwarranted doubts, which can only occasion additional light to become elicited, than to risk by silence letting a question settle to rest, while any unsupported assumptions are involved in it.

If I rightly apprehend Mr. Penn's ideas, they are these:

Secondary limestones were originally in a soft state.

The waters of the deluge while elevated above England, deposited on it a layer, or bed, of "a soft and plastic" calcareous matter.

On their departure from the earth, by flowing away towards the north, they floated over England the carcases of a number of tropical animals, clustered together into great masses.

These masses became buried in the calcareous mud.

On the sinking of the waters of the deluge below the surface of England, the bed of calcareous mud began to dry, and on doing so completely, became the present Kirkdale rock.

The clustered animal bodies enclosed in the calcareous paste, by putrifying, evolved a great quantity of gas, which forced the limestone paste in all directions from them, and thus generated the Cave in which Mr. Buckland found their bones.

Soft State of Secondary Limestones.

That secondary limestones have been in a state to admit foreign bodies into their substance, their existence in it is evidence.

Every shell and stone on the beach tells by its rounded form the attrition to which it is subject at each flood and ebb of the tide; and that a subtil powder is abraded from it which is collected somewhere.

From the immense multitudes of marine bodies which exist in some of these limestones, from others consisting in fact entirely of them, from in general little or nothing but calcareous matter being present, it becomes highly probable that it is to the calcareous part of marine animals, more or less comminuted, that secondary limestones owe their origin.

Deposition of the Calcareous Mud.

The waters of the deluge had not, surely, either a duration or power, to obtain the matter of this supposed layer of mud.

No shores any longer existing, shells could not be pulverized by the beat of the wave, for it is not deep under water that such destruction is effected ; nor, was it so, would the short period of a year have been sufficient to produce the material of all the secondary limestones of the earth ?

To have harrowed up this matter from the depths of the ocean, would have required an agitation of the waters, which nothing warrants us in giving to them, which every thing denies their having had.

No hurricanes, no tempestuous winds, no swollen billows, are recorded. To drown mankind they were superfluous. A wind having arisen at the termination of the calamity tells that none existed before; and this wind must have been a most gentle one, a very zephyr. A vessel, bulky beyond all the efforts of imagination to figure, so laden, so manned, could not have lived in any agitated sea, least in one which out-topped the Alps, and the Andes, all that could curb its fury, and mitigate its violence.

Had the ark not foundered, which is impossible, what yet had become of the millions which its sides enclosed ? Few had survived to repair the effects of the divine wrath.

The waters must have been at rest when the ark continued stationary for many months on the mountains of Ararat.

Nor, do the agitations of a sea extend far below its surface. What navigator has told of the storm in which the sea became thick with its own sediments ?

But had such a deposit been made on our island, it would not have continued on it. Standing like a little turret in the bosom of the waters, each agitation of them would have precipitated part of it down its sides. Their gigantic tides must alone have washed it away, and on the rush of their final departure, not a vestige of it could possibly have remained behind.

If the waters of the deluge placed a bed of calcareous matter on England and Germany, they must have done so over the entire earth. It must have been an universal stratum.

Yet so total was the deficiency of it at Botany Bay, that the first settlers, for the very little lime which a few structures of immediate necessity required, were compelled, though spare as were the hands, and much as they were wanted for other purposes, laboriously and tediously, to collect shells along the beach. Where a limestone nodule was so anxiously sought and could not be found, great strata could not be near.

But the sediment of the deluge waters would not be mere calcareous matter. It must have consisted of everything which they could receive, suspend, and deposit.

If over the earth were spread such a layer of mire, Noah and the animals could not have landed upon it. Or had they not sunk into it and been smothered; where yet had the weak found refuge from the voracious; where had the herbivorous found food ?

What a time must have elapsed before Noah could cultivate the vine ! Nor is it from such a soil that the wine would have intoxicated the holy Patriarch. Had things so been, Ham never had offended, nor Canaan incurred the fatal curse.

Sinking of the Bodies into the Mud.

Supposing, however, such a bed of " soft and plastic " calcareous matter deposited by the waters on England, the immersion of the bodies into it is of no small difficulty.

Animal bodies bloated with gas from decay, which water had " floated on its surface," are not easily conceived to have displaced a stony powder of a specific gravity of 2.7, and to have fallen below it.

" Turbulent vortices," which are imagined to have lent

their aid on the occasion, would have disseminated the clustered animals, and dispersed the powdery stratum.

That the bodies should in every case have descended into the calcareous pulp, in one unbroken group; that in none a fragment, even a lock of hair, should have parted from the putrid mass, and stopped by the way, cannot certainly plead probability in its favour. Yet what cabinet shows even the slenderest bone of a water-rat bedded in the solid stone? What limestone stratum has astonished the learned, by presenting them, in its substance, with an antediluvian hyæna's bristles, or lion's mane?

Formation of the Cave.

If the limestone pulp was too thin, the gas would pass through it and escape, and the pulp fall back upon the bodies; if too thick, the elastic force of the gas would be insufficient to repel it from them. A precise point of induration, at which it would at once yield and resist, was indispensable. This exact condition would but rarely occur; would, at least, often not do it, and consequently bodies buried in the solid rock must be frequent, if not most so.

It is incredible that in every case the gas should have driven away from the bodies the whole of the mud in contact with them. Some of the mud must have insinuated itself between the several individuals of the cluster, some have penetrated by the mouth, by lacerations, into the cavity of the bodies, and isolated pieces of rock must now occur among the bones, bearing the impression of the parts with which they had been in contact; as at Pompeii, indurated ashes presented the cast of a woman's breasts.

As the parts receded from the bodies, it would carry with it some adhering fragments of them—bones, teeth, hair, feathers; and which would now be fixed to the sides and roofs of the caves.

Bodies which had been previously putrefying for twelve months in a tropical temperature, would not probably have still afforded, after their interment, sufficient gas for the supposed purpose. From some experiments, made a great number of years ago, on the decay of animal muscle confined over mercury, I am inclined to believe, that in no case, when secluded from oxygen, is any great volume of gas evolved by it. Subjected to the imagined pressure, would the matters of the gases have been able to expand to the elastic form? Would they not rather have assumed the fluid one?

Under these circumstances, would the muscular part of the bodies have entirely disappeared? Would not some portion of it have altered to adipocire? In such a state some of it must at least some times be met with.

That fish have, in some cases, been inclosed in strata, invested with all their muscular part, seems indubitable, from the presence of the scales; but they are scattered singly through the stratum, and have blown up no caves round themselves.

Indeed, the clustering of the quadrupeds during their voyage, appears to be by no means a certain event. If they sunk below the surface, they would sink to different levels; borne on the surface, they might assemble together, but no adherence would take place between them, and upon the slightest impulse they would part again.

If the bodies were deposited with their integuments, the bones must be nearly all of them entire. How should they have become broken, enveloped in a soft mass, rendered additionally elastic by the gases of a putrefying state, and floating on a sea which, high above all land, bore them out of the reach of every means of concussion, especially become shivered as are of those of the cave? The force which could thus destroy the bones, had reduced the muscular matter to pulp, and the waters had carried it off, and the cave had had no efficient cause.

If the bodies were deposited entire, every bone of each must be forth coming, and its complete skeleton admit of being mounted.

Between "the animal remains discovered buried singly in strata of gravel and clay, and those found in multitudinous masses in the cavities of rocks," there exist the important differences of the former not being in caves, and of the strata in which they occur being fresh-water ones. These remains may consequently be supposed those of animals washed from heights by inundations, and buried in the earthy matter transported with them.

Nor can the bones of the cave be assimilated to the "shells kneaded into the limestone rock of Portland." For the comparison to hold, the bones must be "kneaded into the limestone rock" as the shells are, and as are the bones in the Stunsfield slate, which have been placed in it by the sea.

If the stalactites had been produced by the descent of portions of the calcareous pulpy mass yielding to gravity, they would, like the stalactites of lava, formed in this way, have the texture of the rock. The stalactites of limestone strata are clusters of crystals, which have generated from solution in water.

Induration of the Calcareous Stratum.

The calcareous paste is supposed by Mr. Penn to have petrified by simple drying; and on this supposition much of the hypothesis concerning the formation of the Cave reposes.

Experiments will convince that a paste of calcareous powder and water does not dry to marble, but to whitening. An indurating faculty must not be attributed to time, it has it not. Chalk strata cannot be assigned a less age than the rocks of Yorkshire, and they have not dried to stone, nor seem hastening to become such.

Each particle of powder is a diminutive pebble, and an intervening cement is required to connect it with the neighbouring ones.

Carbonate of lime dissolved in water by means of an

WRITINGS OF JAMES SMITHSON.

excess of acid is the element of agglutination, which nature has in these cases made use of. The acid solvent exhales or becomes saturated, and the neutral salt, ceasing to be soluble, crystallizes on the particles of the powder.

It is thus that the sands of the Calabrian shores are consolidated. The sea water loaded with the calcareous salt, carries it into them. It cannot be by drying since they are wetted by every wave; and sand wetted with ordinary sea water and dried is not converted into millstone. The great hardness is due to the silicious part.

I brought a mass of sand from the sea at Dumbarton, inclosing a recent razor shell with its epidermis on it, and fragments of coal, cemented to stone by carbonate of lime, so that the calabrain process takes place on that coast.

In limestones consisting of considerable-sized fragments of shells, the sparry cement which connects them is perfectly evident. It is this cement which appears as regular crystals where cavities occur in the mass too large to have been filled by it.

Beds of sediment can by this means become rock at the utmost depths of the ocean, and it is in all probability there that most of them have done so. The workings of contiguous volcanos have supplied the carbonic acid.

Oolites have been evidently formed in a sea much loaded with dissolved carbonate of lime, and which on the escape of the dissolving acid has crystallized round floating particles. When the weight of the grains has become such as to occasion their subsidence, they have been cemented together, every thing taking place in all respects as in the case of the pisolites of Carlsbaden. The Kirkdale rock being composed of oolites must have had this origin.

Such a formation cannot be assigned to the time of the deluge. Besides the violence of bringing within the compass of a few months, operations whose accomplishment seems to have required centuries of centuries, the necessary conditions must have been wanting. Had not all the volcanos become extinguished, they could not, and in such a

time, have poured forth carbonic acid to saturate the im-
mensity of its waters; and it is also utterly impossible to
believe that the beings in the ark, already not a little incon-
venienced for respiration, could withstand the suffocating
effluvium.

Coming of the Animals by Sea.

Of the animals having been tropical ones no testimony is
offered. The elephant of Siberia being now ascertained to
have been a very hairy animal may be supposed to have
been a northern one, and if there were formerly northern
elephants, there may have been northern hyænas and northern tigers.

If the bodies were brought by water, no reason appears
why they are, with the exception of a few birds, exclusively
those of quadrupeds. Reptiles, insects, trees, even fish, for
all of them must have perished from the mixture of salt
and fresh water, must have entangled in the clusters.

As the bodies must have been macerated for about a year
in the tropical seas, before the retreat of the waters transported them towards the north, those of the smaller animals,
as the water-rats, must have been so completely decayed as
to be reduced to the bones solely, which water would not
float.

The voyage from the tropics of the balls of album græcum
in an entire state, is what will not, under any circumstances,
be easy to admit; to suppose it amidst "turbulent vortices,
by which the framework of the animals was shattered, dis-
located, fractured within the integuments," reduced to splin-
ters, is utterly impossible. The entire state of the balls of
album græcum, and the extremely fractured one of the
bones, are totally incompatible on Mr. Penn's system. And
such an ablution would not have left in these balls a trace
of the triple phosphate.

But quadrupeds are not the only animals of tropical
features found in northern latitudes. Every shell in the
strata, the nautili, the cornu ammones, the belemnites, the

anomia, are now as foreign to the surrounding seas, as are the others to the land. If one then came from afar, both did.

What must have been the mass and impetuosity of the wave which could buoy a huge oyster, a massive brain stone, from the equator to the British Islands, and at an elevation to deposit it on Shotover Hill, or at Kingsweston? Such waves had tumbled down the mountains of the earth, shivered its islands and its continents, and choked up the bed of the ocean with their ruins. Surely it is a far less difficulty to " bring the climate to the exuviæ, than the exuviæ to the climate."

The existence together of the bones of many species does not necessitate the conclusion of the animals having been associates in the cave. If hyænas " do not always resort to the same den," neither is it probable do other wild beasts. A succession of inhabitants is admissible.

Nor is it required to believe that any of the animals whose bones were found in the cave died there. If hyænas collect bones round their dens, it must be allowed not very improbable that they sometimes, often even, carry them a little further. Alarmed by the roar of a more mighty devourer, or even by that of one of equal strength, it seems natural for them to retreat with their spoil to their last refuge. Why, but to be able to do this, do they bring them near their dens?

The smallness of the cave's mouth, admitting it to have been always what it now is, would indeed oppose the idea of elephants having walked into it, but no entire skeleton requires the admission of their having done this; and hyænas who feed on putrid carcases, may have found no difficulty in parceling such; or they may have collected " the Bushman's harvest," or the bones may have been carried into the cave by animals more powerful than hyænas.

If animals as ravenous of bones as hyænas are said to be did not, in any hour of dearth, devour those of the water-rats, it may be because those became tenants of the cave
8

only when the water had expelled the hyænas. It is alike improbable that animals of such contrary habits should dwell together, and that hyænas should carry so diminutive a prey as a water-rat, to their den to devour it.

The small quantity of the album græcum can afford no argument against the animals who produced it having lived in the cave. So brittle a substance could not last long under the trample of numerous animals of such bulk. The water which subsequently entered the cave may have destroyed a part. The existence of any is a strong circumstance in favour of the supposition of their having lived in the cave, and such as it would scarcely have dared to hope for, in its support.

If bones of quadrupeds are found inclosed in no rocks but limestone ones, which it may, however, require more extended observation to establish, the reason may be, that in no other rocks are caverns, in which wild beasts can take shelter, so common. These are likewise the only rocks in which the formation of stalactite would close the openings, and preserve the bones through a long course of ages, and so as to have reached our times, from the decay and all the accidents to which in an open cave they would be exposed.

Of the Deluge.

Should every argument which has been adduced to establish that the animals were not brought from remote regions by water, that they lived and died in the countries in which their remains now lie, have appeared insufficient for the purpose, yet, that it is not to the Mosaical flood that their existence, where they now are, is to be referred, two great facts appear to place beyond controversy.

One is the total absence in the fossil world of all human remains of every vestige of man himself and of his arts.

The magnitude of the chastisement, the order of nature subverted to produce it, proclaim the multitudes of the criminal. Human bodies by millions must then have covered the waters; they must have formed a material part, if not the principal one, of every group, and human bones be now consequently met with everywhere blended with those of animals.

Objects of human industry and skill must likewise continually occur among the bones. Of the miserable victims of the disaster numbers would be clothed, and have on their persons articles of the most imperishable materials; and the dog would retain his collar, the horse his bit and harness, the ox his yoke. To men who wrought iron and bronze, who manufactured harps and organs, these things must have been familiar.

But more; embalmed within the substance of the diluvian mud, entire cities, with their monuments, with a great part of their inhabitants, with an infinity of things to their use, would remain. Every limestone quarry should daily present us with some of these most precious of all antiquities, before which those of Italy and Egypt would shrink to nothing.

How greatly must we regret that this is not the case, that we must relinquish the delightful hope of some day finding in the body of a calcareous mountain, the city of Enoch built by Cain, at the very origin of the world, with what awful sentiments had not present generations contemplated objects which once had been looked upon by eyes which had seen the divinity!

The other great fact which forcibly militates against the diluvian hypothesis is, that the fossil animals are not those which existed at the time of the deluge. The diluvian species must have been the same as the present. The multifarious wonders of the ark had for sole object their preservation; while of the fossil kinds, not perhaps one, or quadruped, or bird, or fish, or shell, or insect, or plant, is now alive.

" Amazing proofs of inundations at high levels " are appealed to. Had they being, of the deluge they could at most speak but to their existence; on its influence in the

contested cases, they would be silent; but it appears that
this stupendous prodigy,

> " Like the baseless fabric of a vision,
> Left not a wreck behind."

Of the occurrence of marine depositions at great altitudes,
the elevation of the stratum by volcanic efforts, furnishes a
far more easy solution than the elevation of the sea, as it
refers the phenomenon to a natural cause, and does not
require the immediate interposition of the divine hand;
and the ruptured state and erect position of the strata on
all these occasions, testify strongly in favour of the simpler
supposition.

To collate the revered volume with the great book of
nature, and show in their agreement one author to both,
was an undertaking worthy of the union of piety and
science. If the result has not been what was anticipated;
if we look in vain over the face of our globe for those
mighty impressions of an universal deluge, which reason
tells us that it must have produced and left behind itself, to
some cause as out of the natural course of things as was
that event, must this doubtless be attributed.

By his entering into a covenant with man and brute animals, and having for ever " set his bow in the cloud," as a
token that the direful scene should never be renewed, the
Creator appears to have repined at the severity of his
justice.

The spectacle of a desolated world,—of fertility laid
waste,—of the painful works of industry and genius over-
thrown,—of infantine innocence involved in indiscriminate
misery with the hardened offender,—of brute nature whose
want of reason precluded it from the possibility of all
offence, made share in the forfeit of human depravity, may
be supposed to have touched his heart.

Under the impression of these paternal feelings, to obliterate every trace of the dreadful scourge, remove every
remnant of the frightful havoc, seem the natural effects of
his benevolence and power. As a lesson to the races which

were to issue from the loins of the few who had been spared,—races which were to be wicked indeed as those which had preceded them, but which were promised exemption from a like punishment, to have preserved any memento of them would have been useless.

To a miracle then which swept away all that could recall that day of death when " the windows of heaven were opened" upon mankind, must we refer what no natural means are adequate to explain.

A LETTER FROM DR. BLACK DESCRIBING A VERY SENSIBLE BALANCE.

From Thomson's Annals of Philosophy, Vol. XXVI; New Series, Vol. X, 1825, page 52.

EDINBURGH, *September* 18, 1790.

DEAR SIR : I had the pleasure to receive your letter of the 9th. The apparatus I use for weighing small globules of metals, or the like, is as follows : A thin piece of fir wood not thicker than a shilling, and a foot long, $\frac{3}{10}$ of an inch broad in the middle, and $\frac{1}{10}$ at each end, is divided by transverse lines into 20 parts; that is, 10 parts on each side of the middle. These are the principal divisions, and each of them is subdivided into halves and quarters. Across the middle is fixed one of the smallest needles I could procure to serve as an axis, and it is fixed in its place by means of a little sealing wax. The numeration of the divisions is from the middle to each end of the beam. The fulcrum is a bit of plate brass, the middle of which lies flat on my table when I use the balance, and the two ends are bent up to a right angle so as to stand upright. These two ends are ground at the same time on a flat hone, that the extreme surfaces of them may be in the same plane; and their distance is such that the needle when laid across them rests on them at a small distance from the sides of the beam. They rise above the surface of the table only one and a half or two-tenths of an inch, so that the beam is very limited in its play.

The weights I use are one globule of gold, which weighs one grain; and two or three others which weigh one-tenth of a grain each; and also a number of small rings of fine brass wire made in the manner first mentioned by Mr. Lewis, by appending a weight to the wire, and coiling it with the tension of that weight round a thicker brass wire in a close spiral, after which the extremity of the spiral being tied hard with waxed thread, I put the covered wire in a vice, and applying a sharp knife which is struck with a hammer, I cut through a great number of the coils at one stroke, and find them as exactly equal to one another as can be desired. Those I use happen to be the 1-30th part of a grain each, or 300 of them weigh 10 grains; but I have others much lighter.

You will perceive that by means of these weights placed on different parts of the beam, I can learn the weight of any little mass from one grain or a little more to the $\frac{1}{1200}$ of a grain. For if the thing to be weighed weighs one grain, it will, when placed on one extremity of the beam, counterpoise the large gold weight at the other extremity. If it weighs half a grain, it will counterpoise the heavy gold weight placed at 5. If it weigh $\frac{6}{10}$ of a grain, you must place the heavy gold weight at 5, and one of the lighter ones at the extremity to counterpoise it; and if it weighs only 1, or 2, or 3, or 4-100ths of a grain, it will be counterpoised by one of the small gold weights placed at the first, or second, or third, or fourth division. If on the contrary it weigh one grain and a fraction, it will be counterpoised by the heavy gold weight at the extremity, and one or more of the lighter ones placed in some other part of the beam.

This beam has served me hitherto for every purpose; but had I occasion for a more delicate one, I could make it easily by taking a much thinner and lighter slip of wood, and grinding the needle to give it an edge. It would also be easy to make it carry small scales of paper for particular purposes.

We have no chemical news. I am employed in examining the Iceland waters, but have been often interrupted. I never heard before of the quartz-like crystals of barytes aërata, nor of the sand and new earth from New Holland. Indistinct reports of new metals have reached us, but no particulars. Some further account of these things from you will, therefore, be very agreeable. Dr. Hutton joins me in compliments, and wishing you all good things; and I am, Dear Sir,

Your faithful humble servant,
JOSEPH BLACK.

NOTE BY Mr. SMITHSON.—The rings mentioned above have the defect of their weight being entirely accidental; and consequéntly most times very inconvenient fractions of the grain. I have found that a preferable method is to ascertain the weight of a certain length of wire, and then take the length of it which corresponds to the weight wanted. If fine wire is employed, a set of small weights may be thus made with great accuracy and ease. Inconvenience from the length of the wire in the higher weights is obviated by rolling it round a cylindrical body to a ring, and twisting this to a cord.

This little balance is a very valuable addition to the blow-pipe apparatus, as it enables the determination of quantities in the experiments with that instrument, which was an unhoped-for accession to its powers.

Dr. Black mentioned to me its having been used by an

assayer in Cornwall, to whom he had made it known ; and I
have since heard, from another person, of an assayer in that
county, who, finding the assays he was employed to make,
cost him more in fuel than he was paid for them, had con-
trived means of making them at the blowpipe on one grain
of matter. I presume him to have been the same Dr. Black
had spoken of.

LONDON, *May* 12, 1825.

A METHOD OF FIXING CRAYON COLOURS.

From Thomson's Annals of Philosophy, Vol. XXVI; New Series, Vol. X,
1825, page 236.

LONDON, *August* 23, 1825.

GENTLEMEN : Wishing to transport a crayon portrait to a
distance for the sake of the likeness, but without the frame
and glass, which were bulky and heavy, I applied to a man
from whom I expected information for a method of fixing
the colours. He had heard of milk being spread with a
brush over them, but I really did not conceive this process
of sufficient promise to be disposed to make trial of it.

I had myself read of fixing crayon colours by sprinkling
solution of isinglass from a brush upon them, but to this
too, I apprehended the objections of tediousness, of dirty
operation, and perhaps of incomplete result.

On thinking on the subject, the first idea which presented
itself to me was that of gum-water applied to the *back* of
the picture ; but as it was drawn on sized blue paper, pasted
on canvass, there seemed little prospect of this fluid pene-
trating. But an oil would do so, and a drying one would
accomplish my object. I applied ̖drying oil diluted with
spirit of turpentine ; after a day or two when this was grown
dry, I spread a coat of the mixture over the front of the
picture, and my crayon drawing became an oil painting.

NOTES:

AND ADDENDA TO TITLES.

Page 29:

In a critical notice of Davy's Elements of Chemical Philosophy in the Quarterly Review for 1812, the writer speaking of recent advances in chemistry, and especially in the establishment and extension of the law of definite proportions, remarks: "for these facts the science is principally indebted after Mr. Higgins, to Dalton, Gay Lussac, Smithson, and Wollaston." Quarterly Review, 1812, vol. viii, p. 77.

Page 34: On the composition of the compound sulphuret from Huel Boys, and an account of its crystals—otherwise called Bournonite.

Page 42: On the Composition of Zeolite.

This article was translated by Smithson himself into French, and published under the title "Memoire sur la Composition de la Zéolite," in the Journal de Physique, de Chimie, et d'Hist. Nat., etc. Paris, 1814, vol. lxxix, pp. 144–149.

Page 47: On a substance from the Elm Tree, called Ulmin.

This article (translated by M. Vogel) was published under the title "Expériences sur l'Ulmine," in the Journal de Physique, de Chimie et d'Histoire Naturelle. Paris, 1814, vol. lxxviii, pp. 811–815.

Page 65: On a native compound of sulphuret of lead and arsenic.—*Binnit* of Naumann.

Page 68: Thomson's Annals of Philosophy October, 1821, vol. ii, New Series, pp. 291–292. Contains comments by Charles König, on Smithson's article on "Fibrous Metallic Copper."

Page 71: An account of a native combination of sulphate of barium and fluoride of calcium.

Das von Smithson als Flussbaryt aufgeführte Mineral aus Derbyshire ist wohl nur ein sehr inniges Gemenge von Fluorit und Baryt. (Naumann, Min. 9th edit., p. 261, Ann. 3.)

A MEMOIR ON THE SCIENTIFIC CHARACTER AND RESEARCHES OF JAMES SMITHSON, ESQ., F.R.S.,

By WALTER R. JOHNSON,

Corresponding Secretary of the Academy of Natural Sciences of Philadelphia, Member of the National Institute, &c.

Read before the National Institute, Washington, D. C., April 0, 1844.*

PRELIMINARY NOTE.

In the many notices of Mr. Smithson's bequest, and plans for establishing an institution on its basis, which have either officially or otherwise been brought before the public, no succinct account has, so far as the writer's recollection serves, been offered of the scientific pursuits of Mr. Smithson himself,—a very material omission, it is conceived,—and one which could not fail to encourage, or at least excuse, the multiplication of schemes, for carrying out the provisions of his will. A knowledge of the habits, pursuits and feelings of the testator, on the contrary, may relieve us from uncertainty in the interpretation of his language, and the application of his bequest.

If the gratitude of posterity attaches to the memory of successful warriors who enlarge the boundaries of a nation's physical domain, much more is it due to him who opens the fields of knowledge, invites ardent votaries to their cultivation, and thus promotes that nation's happiness, glory, and prosperity.

Under whatever form of government, in whatever social condition, the man of practical benevolence seeks to give his benefactions the character of intellectual blessings; whether, like Bridgewater, he aspires with lofty aim to unravel the designs of creation, explain the final causes of physical laws, and impress by written treatises, the lessons of eternal truth on the matured understandings of men; whether, with the acute, discriminating and practical Girard, he content himself with the humbler but not less honorable office, of rescuing from ignorance, vice and degradation, the homeless and friendless orphan; whether, with Franklin, he found a library; with Maclure endow an academy for researches in natural science; or, with Smithson, seek to stimulate into activity the spirit of philosophical research;

* Philadelphia, Barret & Jones, Printers, 33 Carter's alley, 1844.

123

to " *increase* " by deepening the sources, and " *diffuse* " by multiplying the channels of knowledge ; in each and all of these cases, the universal sentiment of mankind awards a grateful recognition to the intellectual, moral, social benefactor.

But when, in addition to other circumstances of the benefaction, the author has selected for the exercise of his benevolent spirit, not a small circle of votaries of science in a region where the avenues to knowledge are sedulously guarded, but, a great nation, which has made equal rights the basis of its social system, and virtue and intelligence the supports of all its institutions, it is evident that a higher meed of praise, and a deeper feeling of gratitude should spring from the breast of every lover of liberty and of truth.

Having made our country the recipient of his benefaction ; having given us the inheritance of his fame as well as of his fortune, Smithson may justly claim from this side of the Atlantic the tribute of a recognition of his merits, a due appreciation of his own labors, in those paths to which he has invited the scientific efforts of our citizens—efforts on which he has, virtually, and it is to be hoped, not *ineffectually*, invoked the fostering care of this nation's government.

Let one instance in our country suffice—let not a second be exhibited, of that shameful violation of trusts, solemnly assumed, which seeks, in the indulgence of personal vanity, in the execution of splendid schemes of architecture, utterly incongruous to their purpose, or in the search after inapplicable, far-fetched plans of organization, to find a substitute for the simple directions of a man of plain common sense.

On the basis of his labors alone, the true votary of science is willing to rest his credit with mankind, and his fame with future generations. He can look with indifference on the artificial distinctions which fashion, and the greedy love of notoriety, conspire to throw or to draw around pretension and mediocrity. As he deals with the great truths of nature, and not with the changeful humors of man ; as he investigates and promulgates laws, not subject to REPEAL ; announces *results*, not of bargains and compromises, but of the eternal fitness and congruity of parts in creation, he experiences none of the feverish anxiety about adverse interests, that may one day undo his works, which often accompanies the labors of men in other walks of intellectual effort.

In the view of such a man, the accidents of birth, of fortune, of local habitation, and conventional rank in the artificial organization of society, all sink into insignificance by the side of a single truth of nature. If he have contributed his mite to the "*increase*" of knowledge; if he have diffused that knowledge for the benefit of 'man; and, above all, if he have applied it to the useful, or even to the ornamental purposes of life, he has laid not his family, not his country, but the world of mankind under a lasting obligation.

As with societies, so with individuals occupying themselves with scientific pursuits, the estimation in which they must be held, will ever depend on the amount, but especially upon the quality of new published truths which they disseminate. Hence we look primarily to the published works of a scientific man for the evidences of what he has done for science.

They whose recollections of scientific works go back to the first years of the present century, will have no difficulty in judging how far the principle just stated will rank James Smithson among the working scientific men of his time. The transactions of the Royal Society of London, and the scientific journals of the day, will, without reference to other evidence, place us in a condition to solve this question.

But we are fortunately not left to these alone. In his written journals, scientific notes, and more elaborate manuscript papers on a great variety of topics, connected with his tours of observation, and with his studies in numerous departments, we witness the workings of a mind ever active in its endeavors to elicit from the volume of nature truths worthy to fix the attention of all intelligent beings. Let us first recur to his printed works.

1. In the Philosophical Transactions, vol. 93, is a paper *on the Chemical Analysis of some Calamines.* Read November 18, 1802.

In this paper the author describes calamine—1, from Bleyburg in Carinthia; 2, from Somersetshire; 3, from Derbyshire; and 4, electrical calamine.

In this essay the author remarks that " Chemistry is yet so new a science; what we know of it bears so small a proportion to what we are ignorant of; our knowledge in every department of it is so incomplete, consisting so entirely of isolated points, thinly scattered, like lurid specks on a vast field of darkness, that no researches can be undertaken without producing some facts leading to consequences which extend beyond the boundaries of their immediate object."

The Abbe Haüy had advanced the opinion that calamines were all of one species, and all mere oxides or " calces " of zinc, containing no carbonic acid, and that their effervescence with acids was due to an accidental admixture of carbouate of lime. Smithson's analyses completely overthrew this opinion, and established these minerals in the rank of true carbonates.

His remarks on the action of the ores of zinc before the blow-pipe, evince much discernment in relation to the effects there observed.

"The exhalation of these calamines at the blow-pipe, and the flowers which they diffuse round them on the coal, are probably not to be attributed to a direct volatilization of them. It is more probable that they are the consequence of the disoxidation of the zinc calx, by the coal, and the inflammable matter of the flame, its sublimation in a metallic state, and instantaneous recalcination. And this alternate reduction and combustion may explain the peculiar phosphoric appearance by calces of zinc at the blow-pipe."

"The apparent sublimation of the common flowers of zinc at the instant of their production, though totally unsublimable afterwards, is certainly, likewise, but a deceptive appearance. The reguline zinc, vaporized by the heat, rises from the crucible, as a metallic gas, and is, while in this state, converted to calx (oxide.) The flame which attended the process is a proof of it.

"The fibrous form of the flowers of zinc is owing to a crystallization of the calx while in mechanical suspension in the air, like that which takes place with camphor when, after having been sometime inflamed, it is blown out."

As incidental to this inquiry on calamines, he introduces a remark of great interest in connection with the subject of crystallization—a subject, which, when applied to a particular body of the highest interest to the arts, (I refer to wrought iron,) has of late awakened great attention both among practical and scientific inquirers; and which has been invested with a deep tragic interest by a recent lamentable occurrence in our own community :

"A moment's reflection," says Smithson, "must evince how injudicious is the common opinion of crystallization requiring a state of dissolution in the matter, since it must be evident that while solution subsists, as long as a quantity of fluid admitting of it is present, no crystallization can take place. The only requisite for this operation is a freedom of motion in the masses which tend to unite, which allows them to yield to the impulse which propels them together, and to obey that sort of polarity which occasions them to present to each other the parts adapted to mutual union.

"No state so completely affords these conditions as that of mechanical suspension in a fluid, whose density is relatively, to their size, such as to oppose a resistance to their descent in it, and to occasion their mutual attraction to become a power superior to their force of gravitation.

"It is in these circumstances that the atoms of matter find themselves, when, on the separation from them of the portion of fluid by which they were dissolved, they were abandoned in a disengaged state in the bosom of a solution, and hence it is in saturated solutions sustaining evaporation, or equivalent cooling, and free from any perturbing motion, that regular crystallization is usually effected. But those who are familiar with chemical operations, know the sort of agglutination which happens between the particles of subsided and very fine precipitates, occasioning them, on a second diffusion through the fluid, to settle again much more quickly than before, and which is certainly a crystallization, but under circumstances very unfavorable to its perfect performance."

The recent discovery of the reduction of wrought-iron from a fibrous to a granular state by a mechanical percussion, especially at a certain elevated temperature, is a case strongly illustrative of the views of Smithson on this abstruse and difficult subject.

In the same paper (on the calamines) he has attempted to show a simple definite relation to exist between the constituents of this material.

In attestation of the value of these observations by Smithson, we may cite Gregory Watt's paper on the basalts published in the following year, (1803 :)

"It has been most justly remarked by Mr. Smithson, that solution, far from being necessary to crystallization, effectually prevents its commencement; for, while solution subsists, crystallization cannot take place. It may remain a question, whether previous solution be essential as a preparatory means of obtaining by subsequent evaporation, the small parts of bodies disengaged so that they may unite to form regular crystals. If by solution be only meant that simple action of heat or water which merely counteracts the force of aggregation, and relieves the molecules from their bond of union with each other, it certainly *is* a requisite; but if by solution be meant that action of affinities by which not only the force of aggregation is overcome, but the combinations which constitute the molecules are destroyed, it obviously is not only unnecessary, but prejudicial to the crystallization; as a new set of molecules must be formed, by a new combination of the elementary particles, before the formation of regular bodies can take place. The suspension of the molecules ready to crystallize may be correctly said to be merely mechanical. Though the mechanical action of trituration can never be expected to resolve even the most divisible body into its molecules, because the fractures will be at least as frequently across the natural joints as in their direction; yet, even by this rude method, some perfect molecules may be disengaged; for we find that water, passing over large surfaces of silicious sand, finds some molecules of silex in the state proper for aggregation, and even for crystallization. Mechanical suspension in a fluid medium of such density that the crystalline polarity may be enabled to counteract the power of gravity, is with justice considered by Mr. Smithson the only requisite for the formation of crystals.

"The particles of bodies apparently solid must be capable of some internal motion enabling them to arrange themselves according to polarity, while they are still solid and *fixed* as far as they have reference to the ordinary characters of fluidity."

The mode of examining calamines, adopted by Smithson, was to subject them to heat, in order to expel water and carbonic acid, and then to dissolve the residue in sulphuric acid, drying the white vitriol thus produced, and estimating the weight of oxide by that of anhydrous sulphate. This estimation of a metallic oxide in its state of a dry sulphate, enables the chemist to avoid two or three operose and troublesome processes, including filtration, washing and igniting, which ordinarily consume much time, labor, and minute attention.

As the result of his careful inquiry into the truth of the position assumed, it appears, by Haüy, without a sufficient examination, Smithson makes the following statement at the conclusion of his paper :

"No calamine has yet occurred to me which was a real uncombined calx of zinc. If such, as a native product, should ever be met with in any of the still unexplored parts of the earth, or exist among the unscrutinized possessions of any cabinet, it will easily be known by producing a quantity of arid vitriol (anhydrous sulphate) of zinc, exactly double of its own weight; while the hydrate of zinc, should it be found single or uncombined with carbonate, will yield 1.5 times the weight of this arid salt."

2. In the Phil. Transactions, vol. 96, p. 267, 1806, is an "*Account of a discovery of Native Minium*," in a letter from James Smithson, Esq., F. R. S., to the Right Hon. Sir Joseph Banks, K. B., P. R. S. Read April 24, 1806.

This letter is dated at Cassel, in Hesse, March 2d, 1806. He states that he has found minium native in the earth—the gangue, compact carbonate of zinc—with a flaky, crystalline appearance. He gives, in the course of his remarks, the chemical reactions and modes of testing employed to detect its nature.

"This native minium," he remarks, "seems to be produced by the decay of a galena, which I suspect to be itself a secondary production from the metallization of the white carbonate of lead by hepatic gas. This is particularly evident in a specimen of this ore, in one part of which is a cluster of large crystals. Having broken one of these it proved to be converted into minium to a considerable thickness, while its centre is still galena."

I may remark, in confirmation, that the mineral veins of iron, copper, lead, and silver of the United States, afford abundant evidences of the production of "secondary" ores,—such as hydrated peroxides of iron, from the argillaceous carbonates, the protoxide and peroxide, and carbonate of copper, from the yellow sulphuret; the carbonate of lead with its protoxide and peroxide, from galena; this last being the reverse of the order of change conjectured by Smithson. In the silver mines of North Carolina, now worked with considerable activity, the metallic silver is at the outcrop of the veins found mixed with carbonate of lead and of copper, phosphate of lead, with other materials much disintegrated, and offering great facilities for their extraction, while at greater depths, below the reach of atmospheric and other surface influences, the body of ore comes to be almost altogether a mass of galena intermixed with metallic silver.

3. In the Phil. Trans. vol. xcviii., p. 55, (1808,) is a paper by Mr. Smithson, " *On the composition of the compound sulphuret from Huel Boys and an account of its crystals*," p. 8, 1 plate. Read January 28, 1808. In this paper the compound sulphuret of lead, antimony, and copper is described with an account of its chemical properties, and theoretical views of the manner in which proximate elements like these co-exist. He states his belief that all combination is binary, that no substance whatever has more than two proximate or true elements. He makes the mineral to consist of—

Sulphuret of lead	49.7
Sulphuret of antimony	29.6
Sulphuret of copper	20.7
	100.

He gives a figure representing the forms of the crystals and the angles formed by the several faces with each other.

In Tilloch's Magazine, vol. xxix., for 1808, in an account of the proceedings of the Royal Society, we have the following remarks relative to this paper : " December 24, 1807. A paper by Mr. Smithson, on quadruple and binary compounds, particularly the sulphurets was read. The author seemed to doubt the propriety of the distinction, or rather the existence of quadruple compounds ; believed that only two substances could enter as elements in the composition of one body, and contended that in cases of quadruple compounds a new and very different substance was formed, which had very little relation to the radical or elementary principles, of which it was believed to be composed. This opinion he supported by reference to the sulphurets of lead, galena, and of antimony, and to the facts developed by crystallography. In the latter science, he took occasion to correct and confirm some remarks of his in the Transactions for 1804, on different crystals, which he acknowledged have not hitherto been found in nature.

4. In the Phil. Trans. vol. ci., p. 171, for 1811, is a paper " *On the composition of Zeolite*," read Feb. 7. 1811.

In the commencement of this paper the author recognizes the principle that mineral bodies are native chemical compounds, and that it is only by chemical means that their species can be ascertained with any degree of certainty. He found the Zeolite to contain,

Silica	49.0
Alumina	27.0
Soda	17.0
"Ice"	09.5

He calls it a "hydrated silicate of alumina and soda."
In relation to this paper on Zeolites, the following notice is contained in Tilloch's Philosophical Magazine, vol. xxxvii., from January to June, 1811, (p. 152,) under the head of the "Proceedings of the Royal Society:"

"February 7th, Mr. Smithson's paper on Zeolite was read. This ingenious mineralogist having received some specimens of this mineral from Häuy himself, and labelled by his own hand, he deemed it a favorable opportunity of ascertaining if there were any chemical difference between the mesotype of the French crystallographer, and zeolith of Klaproth, as he had previously discovered the existence of soda in all the specimens of zeolite, which are found in these kingdoms, as well as those in Germany. M. Vauquelin analyzed several specimens of zeolite, without discovering any traces of soda, but Mr. Smithson discovered alkali even in the mezotype sent him by M. Häuy, and in every other specimen of zeolite in his possession. From this circumstance he is inclined to prefer the original name of zeolite as given to this mineral by its discoverer Cronsted, to that of mezotype, as given it by Häuy, and considers the distinction between mezotype and natrolith as unsupported by chemical analysis."

5. In the Phil. Trans. vol. ciii. (1813,) p. 256, to 262, is a paper "*On a saline substance from Mount Vesuvius.*" Read July 8, 1813.

This paper gives a chemical quantitive analysis of a compound sulphate of potash.

Sulphate of potash	- -	71.4
Sulphate of soda	- -	18.6
Muriate of soda	- -	04.6
Muriate of ammonia ⎫		
Muriate of copper ⎬	- -	05.4
Muriate of iron ⎭		
		———
		100.0

In the commencement of the paper are some very interesting general views relative to the connection of volcanoes with the theory of geology. One remark is worthy of citation:

"In support of the igneous origin here attributed to the primitive strata, I will observe that not only no crystal imbedded in them, such as quartz, garnet, tourmaline, &c., has ever been seen enclosing drops of water, but that none of the materials of these strata contain water in any state."*

6. In the Phil. Transactions, vol. ciii. p. 64, (1813,) is a paper "*On a substance from the Elm Tree, called Ulmine.*" Read December 10, 1812.

This paper gives an account, 1st. Of ulmine received from Sicily; 2d. Of English ulmine; and 3. Of the sap of the elm tree.

---

* In confirmation of this statement see a late paper by Professor Lewis C. Beck, entitled "Views concerning igneous action," in Silliman's Journal, vol. xlvi., page 337, April, 1844.

The experiments were made to determine the properties and composition of the substance.

7. In the Transaction of the Royal Society, vol. cviii., for 1818, p. 110, are " *A few facts relative to the coloring matters of some vegetables.*" Read December 18, 1817.

The vegetables particularly examined and described in this paper are :

  a Turnsol, (litmus,)
  b The violet.
  • *c* Sugarloaf paper.
  d Black mulberry.
  e The common poppy.
  f Sap green, and
  g Some animal greens.

The above paper is chiefly an account of experiments made for the purpose of testing the chemical characters of the coloring materials of the different substances—an exceedingly interesting branch of inquiry in organic chemistry—scarcely much advanced at this day beyond the point at which Mr. Smithson left it.

From the period of 1818, Mr. Smithson appears to have ceased his contributions to the Transactions of the Royal Society. After this time we find his name most frequently occurring in the Annals of Philosophy, a work too well known to require any remarks upon its scientific character.

8. In this periodical, vol. xiv., 1819, is a letter from Mr. Smithson, dated Paris, May 22, 1819, relative to " plombe gomme," in which he claims the discovery of the composition of that substance for his " illustrious and unfortunate friend, and indeed distant relative the late Smithson Tennant," who he asserts had ascertained that it was a combination of oxide of lead, alumina and water.

He describes the ore, its reactions and modes of reduction. The *alumine* was detected by the usual test of igniting, wetting the whitened mass with nitrate of cobalt, and again igniting producing a blue color.

It decrepitated when heated in a glass tube over a candle, and deposited water in the upper part of the tube, thus proving it to be a *hydrate*.

9. In the Annals, same vol., page 96, is another letter dated Paris, May 19, 1819, (three days before the preceding,) in which he describes *a native sulphuret of lead and arsenic*, found in Upper Valais, in Switzerland, in a granose compound of carbonate of lime and magnesia.

He gives the native characters of the ore, its reactions before the blow-pipe and the action of reagents upon it, particularly of a delicate test of the presence of sulphur, which consisted in placing a minute portion of an insoluble sulphate of baryta formed by treating its solution with chloride of baryum on a very small bit of charcoal, heating it strongly, then dipping it in a drop of water on polished silver, giving to the latter a deep black stain.

Mr. Smithson conducted his researches on a minute scale. The above trials were made with particles little more than visible; the results, however, sufficiently established the nature of the constituent parts. The proportions were necessarily left for inquiries on another scale.

The two preceding subjects are honorably noticed in a historical sketch of improvements in physical science during the year 1819, contained in the 16th vol. of the Annals, (1820,) p. 100.

10. In the same vol. (xvi.) of the Annals, are contained two letters to Dr. Thomson, one dated Paris, March 17th, the other March 24th, 1820.

The former contains a "*View of the probable causes which produce fibrous metallic copper, found both in the ores of copper, and in the slag of copper furnaces.*" Mr. Smithson conceives these fibres to be produced by squeezing metallic copper in a state of fusion into or through pores of the glass, while the latter is cooling and contracting.

11. The latter communication contains *An account of a native combination of sulphuret of barium and fluoride of calcium.* This substance was found in Derbyshire, in close proximity with sulphuret of lead.

He describes with great minuteness the reaction of this substance with tests, and infers that it consists of—

Sulp. of Barium,	-	-	51.5
Fluoride of Calcium,	-	-	48.5

12. In the Annals, vol. xvii., p. 271, is a letter from Mr. Smithson, dated February 17, 1821, in which he describes *capillary metallic tin forced through the pores of cast iron.*

13. In the Annals for August, 1822, vol. xx., p. 127, is an article (Art. v.) *On the detection of very minute quantities of arsenic and mercury.*

In this publication he refers to his paper in the Annals for August, 1819, relative to the compound sulphuret of lead and arsenic.

"If arsenic, or any of its compounds, is fused with the nitrate of potash, arseniate of potash is produced, of which the solution affords a brick red precipitate, with nitrate of silver.

"In cases where any sensible portion of the potash of the nitre has become free, it must be saturated with acetous acid, and the saline mixture dried and redissolved in water.

"So small is the quantity of arsenic required for this mode of trial, that a drop of a solution of oxide of arsenic in water, which at a heat of 54.5 deg., Fahr., contains not above $\frac{1}{10}$ of oxide of arsenic, put to nitrate of potash, in the platina spoon, and fused, affords a considerable quantity of arseniate of silver. Hence, whence no solid particles of oxide of arsenic can be obtained, the presence of it may be established by infusing in water the matter which contains it.

"The degree in which this test is sensible is readily determined.

"With 5 2 grains of silver he obtained 6.4 grains of arseniate of silver; but 0.65 grains of silver was recovered from the liquors, so that the arseniate had been furnished by 4.55 grains of silver. In a second trial, 7.7 grains, of which only 6.8 grains precipitated, yielded 9.5 grains of arseniate. The mean is 140.17 from 100 silver."

Before the invention of the method of subliming a ring of arsenic in a glass tube, and that more recently employed by Marsh, of converting it, by means of hydrogen, into arseniuretted hydrogen, the method of Smithson was among the most delicate in use, and, as a means of obtaining collateral evidence of the presence of arsenic, it still continues to be employed.

With respect to mercury, he remarks:

"All the oxides and saline compounds of *mercury* laid in a drop of marine acid, on gold, with a bit of tin, quickly amalgamate the gold.

"A particle of the corrosive sublimate, or a drop of a solution of it may be thus tried. The addition of marine acid is not required in this case. Quantities of mercury may be rendered evident in this way which could not be so by any other means."

This test for mercury, it may be remarked, still keeps its place among the best evidences of the presence of that metal.

"This method will exhibit the mercury in cinnabar. It must be previously boiled with sulphuric acid, in the platina spoon, to convert it into sulphate."

"Cinnabar heated in a solution of potash, on gold, amalgamates it."

"A most minute quantity of metallic mercury may be discovered, in a powder, by placing it in nitric acid, on gold, drying, and adding muriatic acid and tin."

14. In the same volume (xx.) is, at page 363, a letter to the editor of the Annals, *On some improvements on lamps, particularly referring to the form of the wicks*, the employment of wax as their fuel, and the mode of extinguishing them, by putting sound wax to the wicks, and then blowing out the flame.

"It is to be regretted," remarks the author, "that those who cultivate science, frequently withhold improvements in their apparatus and processes, from which they themselves derive advantage, owing to their not deeming them of sufficient magnitude for publication. When the sole view is to further a pursuit of whose importance to mankind a conviction exists, all that can be so, should be imparted, however small may appear the merit which attaches to it."

On the *fuel* for chemical lamps, he remarks :

" Oil is a disagreeable combustible for small experimental purposes, and more especially when lamps are to be carried in travelling. I have therefore substituted wax for it. I employ a wax lamp for the blow-pipe."

15. In the 21st volume of the Annals, p. 340, is a short article, (Art. II.) "*On the crystalline form of ice,*" dated March 14, 1823.

After referring to several contradictory statements, he remarks :

" Hail is always crystals of ice, more or less regular. When they are sufficiently so to allow their form to be ascertained, and which is generally the case, it is constantly, as far as I have observed, that of two hexagonal pyramids, joined base to base, similar to that of the crystals of oxide of silicium, (or quartz,) and of sulphate of potassium. *One of the pyramids is truncated,* which leads to the idea that ice becomes electrified on a variation of its temperature, like the tourmaline, silicate of zinc, &c."

" The two pyramids appeared to form, by their junction, an angle of about 80°.

" Snow presents, in fact, the same form as hail, but imperfect. Its flakes are skeletons of crystals, having the greatest analogy to certain crystals of alum, white sulphuret of iron, &c., whose faces are wanting, and which consist of edges only."

16. In the same volume of the Annals, (xxi.) p. 359, is a short paper on a *Means of discriminating between the sulphates of barium and strontium.* It is dated April 2d, 1823.

Mr. S. states that when these earths are in a soluble state, (in acids,) the easier process is to put a particle into a drop of marine acid, on a plate of glass, and to let the solution crystallize spontaneously.

The crystals of choride of barium, in rectangular eight-sided plates, are immediately distinguishable from the fibrous crystals of the chloride of strontium.

Another method is suggested, that of blending the mineral in fine powder, with chloride of barium, and fusing the mixture, putting the mass into spirits of wine, and inflaming it while heated, over a lamp, the flame is red if any strontium is present.

17. In the same volume of the Annals, at p. 384, is a paper *On the discovery of acids in mineral substances,* dated April 12, 1823. This paper gives specific directions in regard to—1, Sulphuric; 2, Muriatic; 3, Phosphoric; 4, Boracic; 5, Arsenic; 6, Chromic; 7, Molybdic; 8, Tungstic; 9, Nitric; 10, Carbonic; 11, Silicic acids.

18. In the 22d volume, p. 258, of the Annals of Philoso-

phy, is a short paper *On the discovery of chloride of potassium in the earth.*

This discovery resulted from an examination of a red feruginous mass, containing veins of white crystalline matter, part of a block said to have been thrown from Vesuvius.

It was a spongy lava, with sparse crystals of augite, pyroxene, or hornblende, the white crystalline matter was wholly soluble in water, and when laid on silver with sulphate of copper, gave an intense black stain.

The potash was detected by chloride of platinum and by tartaric acid.

When decomposed by nitric acid, nitrate of potassa was the solid obtained by crystallization.

19. At the 30th page of the same volume (xxii.) of the Annals of Philosophy, is a short tract *"On the improved method of making coffee."*

The object is to preserve the aroma of the coffee during the coction, which Mr. Smithson effected in a phial closed with a cork, left loose at first, to allow the escape of air, and afterwards closed tight, and kept immersed in boiling water until the process was concluded. It may, when cooled, be filtered, without losing the aroma, and then returned to the close vessel to be re-heated.

" In all cases means of economy tend to augment and diffuse comfort and happiness. They bring within the reach of the many, what wasteful proceeding confines to the few. By diminishing expenditure on one article, they allow of some other enjoyment which was before unattainable. A reduction in quantity permits an indulgence in superior quality. In the present instance the importance of economy is particularly great, since it is applied to matters of high price, which constitute one of the daily meals of a large portion of the population of the earth.

" That in cookery also, the power of subjecting for an indefinite duration, to a boiling heat, without the slightest dependiture of volatile matter will admit of a beneficial application, is unquestionable."

20. In the same volume of the Annals, (xxii.,) p. 412, is a paper, by Mr. Smithson, *On a method of fixing particles upon the sappare,* (cyanite,) dated October 24, 1823.

He refers to the uncertainty of *physical qualities* to determine the species of minerals. Werner was unable, by this means, to discover the identity of the jargon, (zircon,) and the hyacinth; of the corundum and the sapphire; of his apatite and his spargelstein, and "while he thus parted beings from themselves, as it were, he forced others together, which had nothing in common."

Hence, Smithson infers the necessity of chemical analysis; and, to avoid waste, the practice of analyzing on a very small scale.

To fix the particles of minerals on a sappare, in order to subject them to high temperature, Mr. Smithson employed water with gum, as used by Saussure, who invented the method, but he added refractory clay. The particle of mineral was then made to adhere to this clay, a small portion of it being for this purpose taken upon the end of a flattened platina wire.

21. In the 23d volume of the Annals, (p. 100,) we find a paper, by Mr. Smithson, dated, January 2d, 1824, " *On some compounds of fluorine.*"

In this, he makes the apposite and just remark: that, " a want of due conviction that the materials of the globe, and the products of the laboratory are the same, that what nature affords spontaneously to men, and what the art of the chemist prepares, differ in no ways but in the sources from whence they are derived, has given to the industry of the collector of mineral bodies, an erroneous direction."

" What is essential to a knowledge of chemical beings, has been left in neglect; accidents of small import, often of none, have fixed attention— have engrossed it—and a fertile field of discovery has thus remained *unexplored*, where, otherwise, it would have been exhausted."

His method of illustrating the character of fluor spar, was by fixing with clay a small piece, on a bit of platinum foil, and holding the latter on a clay support, in the end of a bit of glass tube, and thus subjecting it to the action of the blow-pipe.

The topaz was also assayed, and gave out fluorine or fluoric acid. Smithson expresses his conviction that topaz is a compound of fluate of silicium and fluate of alumina.*

He also examined kryolite, which had been observed to diminish in fusibility during fusion.

The result of his experiments were: 1st. That fluorides are in general decomposable by heat, and hence, that " we now have a method of discovering the presence of fluorine." 2d. The theory of these decompositions may be obtained by experiment.

Referring to the minute blow-pipe experiments with which his results had been obtained, he significantly remarks:

" There may be persons, who, measuring the importance of the subject by the magnitude of the object, will cast a supercilious look on this discussion; but the particle and the planet are subject to the same laws, and what is learned upon the one will be known of the other."

22. In the same volume (xxiii.) of the Annals, p. 115, is a short paper of the same date, (January 2, 1824,) containing *An account of an examination of some Egyptian colors.*

---

* At this day he would probably have substituted the terms *fluoride of silicium* and *fluoride of aluminum.*

" More than commonly incurious must he be, who would not find delight in stemming the stream of ages, returning to times long past, and beholding the then existing state of things and of men.

" In the arts of an ancient people, much may be seen concerning them, the progress they had made in knowledge of various kinds, their habits and their ideas on various subjects. Products of skill may likewise occur, either wholly unknown to us, or superior to those which now supply them."

He received from Mr. Curtin, who traveled in Egypt, with Belzoni, a small fragment of the tomb of King Psammis.

It was sculptured in basso relievo and painted, the colors being white, red, black and blue. The white was found to be carbonate of lime; the red, oxide of iron; the black, pounded wood charcoal, the texture of the larger particles being perfectly discernible with a lens, after dissolving out the other coloring matters. The blue was a smalt of glass powder, its tinging matter, however, was not cobalt, but copper. Melted with borax and tin, the red oxide of copper immediately appeared.

23. In the 24th volume of the Annals, p. 50, is a paper of ten pages, bearing date June 10, 1824, and containing *Some observations on Mr. Penn's theory concerning the formation of the Kirkdale cave.*

The writer whose work Smithson criticises, had attempted to account for the bones by referring them to the period of *" the Deluge."* This opinion Mr. Smithson very successfully combats. A confutation is, however, hardly needed by geologists in our day. It is not therefore deemed necessary to follow the writer through the steps of his reasoning.

24. In the 25th volume of the Annals, is a letter from Dr. Black, to Mr. Smithson, describing his delicate balance for weighing minute quantities of metals, and other results of analysis, consisting of a thin bit of fir, with a fine cambric needle for an axis, and an upturned bit of brass for a support. To this apparatus Mr. Smithson suggested some improvement in the formation of the weights.

There is much reason to suppose that the foregoing list of twenty-four papers, does not embrace all the published works of Mr. Smithson. The numerous lists of loci or topics, evidently designed to form the heads of essays or treatises, either found disconnected, or united with loose notes, on each topic, or wrought out into formal essays, of which several are found among his manuscripts, afford ground for believing that he was a contributor to some of the literary productions of the day; but as such pieces generally appear anonymously, it is not easy to ascertain the precise object for which these numerous tracts were composed.

It appears from all which has been cited, from the published works of Smithson, that his was not the character of a mere amateur of science. He was an active and industrious laborer in the most interesting and important branch of research—mineral chemistry.

A contemporary of Davy, and of Wollaston, and a correspondent of Black, Banks, Thomson, and a host of other names renowned in the annals of science, it is evident that his labors had to undergo the scrutiny of those who could easily have detected errors, had any of a serious character been committed.

His was a capacity by no means contemptible, for the operations and expedients of the laboratory. He felt the importance of every help afforded by a simplification of methods and means of research, and the use of minute quantities, and accurate determinations in conducting his inquiries. Many of those " lurid spots in the vast field of darkness," of which he spoke so feelingly, have, since his days of activity, expanded into broad sheets of light. Chemistry has assumed its rank among the exact sciences. Methods and instruments of analysis, unknown to the age of Smithson, have come into familiar use among chemists. These may have rendered less available for the present purposes of science, than they otherwise might have been, a portion of the analysis and other researches of our author. The same may, however, be said of nearly every other writer of his day.

Having dwelt so long on the published papers of Mr. Smithson, it will be practicable to give but a brief account of his unpublished memoirs and other writings. These are comprised in about two hundred manuscripts, besides numberless scraps and miscellaneous notes of a cyclopedical character. Many of these are connected with general subjects of history—the arts—language—rural pursuits—gardening—the construction of buildings, and kindred topics, such as are likely to occupy the thoughts and to constitute the reading of a gentleman of extensive acquirements and liberal views, derived from a long and intimate acquaintance with the world.

In a pretty copious mass of notes on the subject of *habitations*, for example, the materials are discussed under the several heads of situation, exposure, exterior, and interior arrangements, materials for their construction ; contents of rooms, furniture, pictures, statuary, and other objects of taste.

In a tract upon knowledge, he takes occasion to remark, that men may consider themselves as having four sources of knowledge : 1st. Observing. 2d. Reasoning. 3d. Information. 4th. Conjecture. It is evident that in his own acquirements in knowledge, he followed this order of proceding, and did not, as many have done, both before and since his time, begin with conjecturing, proceed next, to ask *information* as to the *opinions* of others, receiving, as sound, all those which tally with the *conjecture*, and rejecting the rest, and end with attempting to *reason* themselves into a belief that this mass of crude fantasies constitutes philosophy. Smithson began the process of acquisition by *observing*. For this purpose he made a number of tours or scientific journeys, taking, as opportunity offered, careful observations of all interesting facts.

It was in 1784, (now sixty years since,) that, in company with Mr. Thornton, Monsieur Faujas De St. Fond, the celebrated French philosopher, and the Count Andrioni, he made one of these tours, through New Castle, Edinburg, Glasgow, Dunbarton, Tarbet, Inverary, Oban, Aross, Turtusk, and the island of Staffa. In all these places observations on the evidences of geological structure, on the mineral contents of rocks, on the superposition of beds, on the methods of mining, smelting ores, and conducting manufacturing processes, were made with all the minuteness which the arrangements of the journey could permit.

The period of two generations of men elapsed since the journey to Fingal's cave was undertaken, has seen a vast accession of strength to that ruling passion which now sends forth the votaries of geology of all countries, with hammer and knapsack, to explore alike the desert and the fertile field, to indulge in the luxury of toilsome wanderings, soiled apparel, hard lodgings, and scanty fare.

The hardships and privations of such expeditions were, at that day, not so often encountered as at present, because the expeditions themselves were seldom undertaken. Still, it would, even in our own time, be thought a very respectable piece of hardihood and scientific self-denial, to encounter such risks and privations as are here and there jotted down in Smithson's journal, in relation to this visit to the island of Staffa.

The party had arrived at a house on the coast of Mull, opposite the island. The journal proceeds :

"Mr. Turtusk got me a separate boat,—set off about half-past eleven o'clock in the morning, on Friday, the 24th of September, for Staffa. Some wind, the sea a little rough,—wind increased, sea ran very high,—rowed round some part of the island, but found it impossible to go before Fingal's cave,—was obliged to return,—landed on Staffa with difficulty,—sailors press to go off again immediately,—am unwilling to depart without having thoroughly examined the island. Resolve to stay all night. Mr. Maclaire stays with me; the other party which was there had already come to the very same determination,—all crammed into one bad hut, though nine of ourselves, besides the family;—supped upon eggs, potatoes, and milk,—lay upon hay, in a kind of barn." (The party, be it remembered, embraced two English gentlemen, one French savan, one Italian count.) "25th. Got up early, sea ran very high, wind extremely strong—no boat could put off. Breakfasted on boiled potatoes and milk; dined upon the same; only got a few very bad fish; supped on potatoes and milk;—lay in the barn, firmly expecting to stay there for a week, without even bread."

"Sunday, the 26th. The man of the island came at five or six o'clock in the morning, to tell us that the wind was dropped, and that it was a good day. Set off in the small boat, which took water so fast that my servant was obliged to bail constantly—the sail, an old plaid—the ropes, old garters."

With this unpromising outfit, however, the party, at length, once more, reach terra firma.

On the 29th, the tourists are at Oban, where a little circumstance is noted, which significantly marks the zeal and activity of the collector of minerals and fossils, and the light in which that devotion to geology is sometimes viewed by the unscientific part of the community:

"September 29. This day packed up my fossils in a barrel, and paid 2s. 6d. for their going by water to Edinburg. Mr. Stevenson charged half a crown a night for my rooms, because I had brought 'stones and dirt,' as he said, into it."

A month later we find him at Northwich:

"October 28. Went to visit one of the salt mines, in which they told me there were two kinds of salt. They let me down in a bucket, in which I only put one foot, and I had a miner with me. I think the first shaft was about thirty yards, at the bottom of which was a pool of water, but on one side there was a horizontal opening, from which sunk a second shaft, which went to the bottom of the pit, and a man let us down in a bucket smaller than the first."

In these trivial incidents we may note the character of an enthusiast in pursuit of his favorite objects; a man not to be turned aside by the fear of a little personal inconvenience from the attainment of his ends. In his tours on the continent, of which, one was made from Geneva to Italy, through Tyrol, in 1792; one through certain parts of Germany, in 1805; another in 1808, and a third from Berlin to Hamburg, in 1809, are found many interesting remarks on the physical features, geology, and climate of the districts of country through which he passed.

What has now been presented, may perhaps enable us to judge of the *animus* which impelled Smithson to found an institution "for the increase and diffusion of knowledge among men."

It may at least enable us to decide whether it was any undue assumption on his part, to constitute himself a patron of science. Those who look at the matter in the humble light of a mere pecuniary transaction, will, readily enough, answer the question. They will say "every man has a right to do what he will with his own."

But the inquiry is one of far higher import, it addresses itself to men of science. *Had Smithson the qualifications which should authorize and require us to defer to his judgment, were he now living, in regard to the specific objects of an institution, founded in the broad and comprehensive terms employed in his will?* To this, I think, there can be but one answer. If anybody has a right to direct how institutions of science should be founded and conducted, it is they, who have inured their own hands to the work, who have taken the laboring oar, and won, by its use, an honorable distinction. Such a man, we have seen, was JAMES SMITHSON.

A single question more.—What would have been the purposes of an institution founded by Smithson in his lifetime?

To this, his *lifetime* is a sufficient answer. ·

Researches to "increase" positive knowledge, and publications to "diffuse" and make that knowledge available to mankind—such were the great objects of his own constant, praiseworthy, and laborious efforts.*

---

* The Smithson fund in possession of the Government of the United States, now amounts (April 10, 1844) to $700,000, of which the interest is $42,000 per annum. Two years' interest are *said* to be unpaid.(?)

ON THE WORKS AND CHARACTER OF JAMES SMITHSON.

By J. R. McD. Irby.*

It is the characteristic of modern biography that it seeks to know the personalities of men. It has ceased altogether to be a mere chronology. It attempts to introduce to us its subjects as we would have known them in actual life and to make them the people of our inward world.

Who that has known the splendid benefits derived from Smithson's great foundation has not felt a desire to know more nearly him from whom the gift proceeded? Who has not been impressed with his persevering philanthropy, when, failing to accomplish his object through the Royal Society of Great Britain, he turned his face to the New World and laid up his name in the new order of things and men? Who has not discerned in this the spirit of a real benefactor of mankind, and not that of a vain builder of his own monument. It is my pleasant task to show something of his way, his work, and his thought.

Smithson's actual additions to knowledge are not great, but they are distinct. It was his misfortune to work at two sciences, chemistry and mineralogy, which were yet in their infancy, and at a third, geology, which, though pregnant to the birth, was still unborn, in a true sense. In the dark beginnings of things, when both ideas and methods are imperfect, it is seldom that the bewildered gropings of men become valued heirlooms to posterity.

We could wish Smithson's name to have been coupled with some great discovery, or with the apprehension of some far-reaching law that would have formed a worthy inscription for the portal of his institution. Though this be not gratified, we shall find that he appreciated the great problems before him and attempted their solution; that he knocked earnestly and worthily at the portal of great knowledge, and that although it was denied him to be the first to enter into the greater chambers, he was, nevertheless, no unworthy seeker. When we have caught the utterances of

* Prepared at the request of the Institution, September, 1878.

143

his writings we shall learn that his mind was tuned to great things.

The greater part of Smithson's work was in analytical chemistry. He discovered several tests, the most important of which is the blow-pipe test for sulphur by reducing its compounds on charcoal with carbonate of sodium, and observing the stain on silver when the fused mass is laid upon it in a drop of water (p. 66). In the paper "On the Detection of very Minute Quantities of Arsenic and Mercury," (p. 75,) two very good tests for these elements are given, especially that for the first:

"If arsenic, or any of its compounds, is fused with nitrate of potash, arseniate of potash is produced, of which the solution affords a brick-red precipitate with nitrate of silver."

The paper on page 82 gives a systematic course for distinguishing the mineral acids. On page 82 a flame-test is given for strontium, which is perhaps the earliest application of colored flames in analytical chemistry. In the paper, page 94, "On some Compounds of Fluorine," the method of detecting this element is described, and a very neat form of apparatus given. The latter is peculiarly convenient in that the etching of glass and the change of color of logwood paper may be simultaneously observed.

A glance through his papers will show how much of his work was actual analysis. Owing to the great improvements in analytical chemistry since his day, his quantitative results are of little value to us. This is not true, though, of the qualitative work. The composition of the so-called Tabasheer (hydrous silica), of the Egyptian colors, the presence of some carbonate in certain calamines, as well as other of his results, have a permanent value. We are apt to overlook them because they are become so obvious and elementary.

Connected with and occasioned by certain of his analyses are some considerations on the laws of the chemical composition of bodies. These, though erroneous, are the greatest of his scientific attempts. They are found on page 27, "Observations" appended to the paper on calamines. These were published in 1802. A further development of his views is found in the paper, page 34, "On the Composition of a Compound Sulphuret from Huel Boys," published in 1808. His idea was that the weights of the proximate constituents of any complex compound bore a simple relation to one another. His experiments lead him to infer that sulphate of zinc is composed of equal weights of ZnO and SO^3. This, though very nearly, is not accurately true; so

nearly that the analytical chemistry of that day was power-
less to detect the difference. His analyses of the Mendip
Hill calamine seemed to show (and did show as nearly as
they showed the truth) that it was composed of— ,

Carbonic acid _____ ½
Calx of zinc _____ ½

He thought to have found further confirmation of his
views in the analysis of the compound sulphuret from Huel
Boys. It must be borne in mind that these attempts were
anterior to the publication of Dalton's theory, (his *Chemical
Philosophy*, appeared in 1808.) The second of the above
mentioned papers was also in 1808, but in the very beginning of the year. He seems to have been absent from England, for he mentions in the beginning of the paper that the
Philosophical Transactions for 1804, had just come into his
hands; and on page 39, paragraph 2, that certain of his
notes were in England. We may be sure he had no knowledge of Dalton's theory. In the paper "On the Composition
of Zeolite," published in 1811, he does not recur to them. I
think these views are worthy of notice in the history of
chemical theory. They were as certainly established as was
possible with the analytical methods of that day.

His very correct apprehension of the true problem of analytical chemistry probably confirmed him in his error. In
the second paper above referred to, on page 35, we find the
following passage :

"We have no real knowledge of the nature of a compound substance
until we are acquainted with its proximate elements, or those matters by
whose direct or immediate union it is produced; for these only are its true
elements. Thus, though we know that vegetable acids consist of oxygene,
hydrogene, and carbon, we are not really acquainted with their composition, because these are not their proximate—that is, their true elements,
but are elements of their elements, or elements of these. It is evident
what would be our acquaintance with sulphate of iron, for example, did
we only know that a crystal of it consisted of iron, sulphur, oxygene and
hydrogene; or of carbonate of lime, if only that it was a compound of
lime, carbon or diamond, and oxygene. In fact, totally dissimilar substances may have the same ultimate elements, and even probably in precisely the same proportions; nitrate of ammonia and hydrate of ammonia
or crystals of caustic volatile alkali both ultimately consist of oxygene,
hydrogene, and azote."

This remarkably lucid passage could not be improved
upon now, three quarters of a century later. Without doubt
his exceedingly clear conception of importance of proximate analysis led him to seek the laws relative to compounds
in their proximate constituents; and he thought to have
found them.

The following passage, page 37, relating to the same

10

subject, shows his perfect understanding of the inductive method, and the inherent indeterminateness of his analysis :

"It is evident there must be a precise quantity in which the elements of compounds are united together in them, otherwise a matter, which was not a simple one, would be liable in its several masses, to vary from itself, according as one or the other of its ingredients chanced to predominate; but chemical experiments are unavoidably attended with too many sources of fallacy for this precise quantity to be discovered by them; it is therefore to theory that we must owe the knowledge of it. For this purpose an hypothesis must be made, and its justness tried by a strict comparison with facts. If they are found at variance, the assumed hypothesis must be relinquished with candor as erroneous; but should it, on the contrary, prove, on a multitude of trials, invariably to accord with the results of observation, as nearly as our means of determination authorize us to expect, we are warranted in believing that the principle of nature is obtained, as we then have all the proofs of its being so, which men can have of the justness of their theories : a constant and perfect agreement with the phenomena, as far as can be discovered."

The following passage, page 29, shows how clearly the object to be attained was set forth in his own mind :

"If the theory here advanced has any foundation in truth the discovery will introduce a degree of rigorous accuracy and certainty into chemistry, of which this science was thought to be ever incapable, by enabling the chemist, like the geometrician, to rectify by calculation the unavoidable errors of his manual operations, and by authorizing him to eliminate from the essential elements of a compound those products of its analysis whose quantity cannot be reduced to any admissible proportion.

"A certain knowledge of the exact proportions of the constituent principles of bodies, may likewise open to our view harmonious analogies between the constitutions of related objects, general laws, &c., which at present totally escape us. In short, if it is founded in truth, its enabling the application of mathematics to chemistry cannot but be productive of material results."

At the time his paper on the "Compounds of Fluorine" was published, the composition of fluor spar was still a matter of doubt. The following is a sketch of a proposed method for determining it :

"If fluor spar, for instance, is a combination of oxide of calcium and fluoric acid, and this is expelled from the oxide merely by the force of fire, the decomposition of it will take place in closed vessels without the presence of oxygen or of water ; fluoric acid will be obtained ; and the weight of this acid and the lime will be equal together to that of the original spar.

"If the spar is metallic calcium and fluorine, and when heated in oxygen absorbs this, and parts with fluorine, it is fluorine which will be collected in the vessels, and its weight and that of the lime will together exceed that of the spar by the oxygen of the lime."

Further on he suggests the employment of vessels of fluor spar for the examination of fluorine. He then discusses the phenomenon of intumescence as observed in fluor spar and similar substances, in order to correct an erroneous explanation of its nature that it was a "new state of equilibrium induced by heat between the constituent parts of a body."

"Why is the change of quality limited to the surface; how has been produced the central cavity; what has forced away the matter which occupied it? A new element has been received from without, one which existed in the matter has been parted with in a state of vapor. This double action may probably be inferred wherever a matter presents this species of vegetation," (p. 100.)

As the story of his analysis of a tear indicates, he was an exceedingly nice manipulator. He was one of the very first who commenced the cardinal practice of modern analytical chemistry, the use of delicate methods and small quantities of material. His quantitative determinations were usually made with about a gramme, and his qualitative determinations often with almost invisible bits. In the examination of the "Native Compound of Sulphuret of Lead and Arsenic" (binnite of Naumann) from Upper Valois, his "trials were made with particles little more than visible." On page 95 he says: "A very minute fragment of fluor spar is fastened by means of clay to the end of a platina wire nearly as fine as a hair, which is the size I now employ even with fluxes." We have before noticed the neat and simple apparatus (p. 97) for the detection of fluorine. On page 86 a method of making and using thin clay plates is given, which might, at the present time, be advantageously employed in blowpipe work, especially if they were made from a pure kaoline. The paper on the "Method of Fixing Particles on the Sappare" (fibres of cyanite) contains repeated instances of his delicacy and neatness.

Smithson's contributions to mineralogy consists principally in the discovery of several new species. Native red lead was first examined by him and its having been derived from galena demonstrated. He also first observed chloride of potassium, in a native state from Vesuvius. He attributed its presence in lava to sublimation. The native compound of sulphuret of lead and arsenic is the rhombic mineral binnite (of Naumann), as is recognized by its locality, chemical composition, hardness and cleavage. He also described a native compound of sulphate of barium and fluoride of calcium from Derbyshire. Naumann (Min., 9te Aufl., p. 261, Anmerkung 3) thinks, as is correct, that this is only a mixture and not a true species.

The crystallographical observations of Smithson are of rather a rough character, owing perhaps to his instruments. They refer to the forms of electric calamine, of bournonite (the compound sulphuret from Huel Boys) and of ice. The

rhombic character of bournonite escaped him, he having
taken it for quadratic. Snow he found to have the form of
a double six-sided pyramid, with a lateral angle of about
80° The various observations on its forms are so discrep-
ant, however, that it is impossible to state which are correct.
On page 81 he gives a crystallographical test to distinguish
between the chlorides of barium and of strontium. The
crystals of the one are rectangular, eight-sided plates; those
of the other fibrous.

At this point, a handful of Smithson's manuscripts may
be mentioned, which escaped the fire at the institution in
1865. They consisted of notes on various specimens of
minerals and rocks belonging to his collection, and also
several fragments of catalogues, which seem to have been
begun in various years. The earliest bears the date 1796,
the latest 1822. These are of little or no scientific value,
except in so far as they illustrate the way in which he
worked. The following are a few extracts from them :

No. 1.—Carbonate of lime containing manganese, from near
   Aix la Chappelle.

It dissolved in nitric acid with effervesence like carbonate of lime. The
salt obtained from this solution by drying over a candle is quite white, but
on heating more it becomes brown, and then on solution in water leaves a
small quantity of brown powder. Prussiate of soda and iron caused a
white precipitate in solution of this stone, and in it the least blue was per-
ceivable. Tincture of galls produced no black color with it.
Some of the above nitrous salt melted on platinum with nitrate of potash gave the green color of manganese.
Copperas put into some of this nitrous solution caused a precipitate of
sulphate of lime.
This carbonate of lime and manganese becomes brown at the blow-pipe.
This carbonate of lime and manganese colored borax red.

No. 19.—Reduced nickel free from arsenic.

It was made at the blow-pipe from oxide of nickel which had been fused
with saltpetre. It contains admixed borax. It is infusible. It probably
contains cobalt.

No. 4.—Crust from the church bell of Torre del Greco,
   formed by the lava in 1794.

There is a crystal in the little group which is the most regular. The two
larger faces of this crystal seem to form an angle of 140° with the prism,
and meet together at the summit in an angle of 80°. There is a broken
crystal in the same group which seems to show that the four larger faces of
the prism form together angles of 90°. The form of these crystals is 8-
sided prisms and 4-sided pyramids and are similar to III. 55. d., having
the four edges of the prism slightly truncated.

No. 7.—A small group of native gold in 24-sided crystals
   from Vöröspatak, in Transylvania.

The matrix is evidently a quartzose stone. Shows in many parts minute crystals of quartz, and contains pyrites disseminated in it, which are probably auriferous.

No. 25.—Arseniate of iron. Paris, September 25, 1820.

1. Nitric acid was put on to some native arsenuret of cobalt to form nitrate of cobalt, and this matter formed as a sulphur colored powder in the mixture. It was washed and dried.

2. Heated in a tube some water and crystals of arsenious acid sublimed and a dark mass remained.

3. This dark mass heated on coal at the blow-pipe emitted fumes probably of arsenious acid and became like a scoria of iron, but the magnet did not effect it.

4. The scoria-like mass dissolved in borax with effervesence and spread much on the coal. This glass in the whole looked black, but where there were air-bubbles it had the color of chrysoberyl.

5. This borax was heated in dilute muriatic acid in a tube. The acid quickly became yellow.

Prussiate of soda and iron formed an abundant precipitate of prussian blue; but nitrate of silver formed only a white curdy precipitate of chloride of silver, and no arseniate of silver.

It is probable, however, that the above yellow powder is arseniate or arsenite of iron.

No. 955.—Paris, May, 1819.

1. In Mr. Stockhausen's catalogue this is called mountain cork, and said to be from Dauphenée.

Both the black fibrous part and the white part, when held in the flame of a candle, take fire and burn with a large flame.

When the white part was tried, a fluid matter like oil flowed from it and ran along the lips of the pincers, and on cooling set with a crystalline texture. The color was greenish, and it was soft and brittle like spermacite.

No fœtid animal smell was perceived during the combustion.

The matter is more like adiposcere than mountain cork.

No. 1166. Octahedral crystals from Clausthal.

1. These crystals are easily broken.

2. Put into pure muriatic acid, the fragments of it did not suffer any change.

3. Per se, at the blow-pipe they did not decrepitate, but readily reduced to a white metal, which exhaled.

4. They dissolve in borax with effervescence, without coloring it. Balls of a white metal were produced, but when the borax became fluid it soaked into the charcoal like alkali, and the whole disappeared.

5. The form of the crystals is regular octahedral, with the six points cut off.

6. Their color is gray, and their aspect metallic.

7. Their fracture is perfectly tubular and parallel to the six corners of the octahedron. Their true form is a cube, fissile, parallel to its six faces.

N. B.—These are, most probably, common sulphuret of lead.

No. 1564.—Native gold from the Edder, a river in Hessia, in Germany.

I had it from Capt. Stockhausen's cabinet.

N. B.—It is only mica.

No. 1639.—A button, which is a white compound of copper, etc.

1. Melted on a bit of slate with saltpetre—the solution of this salt gave a yellow precipitate, with nitrate of silver.

2. Melted on the coal the metal spread; no flowers of oxide. It was very fusible; seemed white while melted; the cooled button filed was yellow like brass; hence, perhaps, an alloy of copper and zinc or tin.

3. It dissolved wholly in nitric acid, forming a clear blue solution; exhaled dry, and pure water added, a small quantity of grey powder was left insoluble.

This solution poured into much water became milky, and some of this milky liquor put into a watch glass with ammonia, and then nitrate of silver added, yielded a yellow precipitate.

No. 1672.—Braunkohle mit Stockwerk vom Ahlberge bei Mariendorf.

In the fire it emits a copious pungent smoke, which pains the eyes greatly. An incombustible residuum remains of the form and nearly size of the bit of wood, which very slowly burned to a white ash. (Paris, March 2, 1820.)

With saltpetre this incombustible residuum burned like anthracite. While the saltpetre was fluid it looked like a dark green color, though not like manganese. On fusing again this color vanished, but on sudden cooling in water the blue was restored. The solution in water was not green, and did not become red.

No. 1766.—Fuller's earth.

1. Does not lose its black color in water.
2. Decrepitates.
3. Melts easily into a black glass, which seems opaque.
4. This black glass is taken up by the magnet.
5. Adheres to the tongue.
6. Rubbed with a little water on a bit of unglazed china it gives a yellow greenish color.
7. Found in a basalt quarry at Wilhelmshöhe, 1804.

No. 2012.—Green clay found by self near Frankfort, April the —, 1805.

1. In a moist state it is very lubric.
2. Compressed in this state to a thin plate it is considerably hard.
3. In the fire hardens and melts to a black glass; is not very fusible, and shows no inflation.
4. Seems to dissolve in borax without much difficulty, and colors it very green. If a great quantity of the clay is put to the borax, a black bead is obtained.
5. I found it adhering to (coating one side of) a mass of lava lately extracted from the earth. It had probably formed in a fissure of the lava stratum.
6. Strongly heated on coal it became black, and the edges melted to a black glass. In this state it was not drawn on by the horse-shoe magnet; but reduced to powder, on a brass plate, some of the powder was taken up.
7. Sulphate of soda and iron did not dissolve it, but the bead became slightly milky on cooling.
8. Put into water it falls into lumps like curds, but which pressed with the fingers, reduce to a powder.

No. 2952. Unknown plated metallic ore, said to come from the Hartz, in my cabinet marked No. 2952.

1. Its color is grey, like that of lead or sulphuret of zinc.
2. It is brittle.
3. Has the metallic gloss and opacity.
4. *Per se* on the charcoal decrepitates greatly.
5. With borax melts, effervesces, emits a white smoke, and exhales, leaving a small ball of white metal, which appears to be lead, as it is entirely fluid when not very hot.
6. Melted in the gold spoon with carbonate of soda produces a greyish mass; water added formed a black powder, and the solution stained silver only very slightly. This solution being mixed with nitric acid produced but a very slight smell of sulphide of soda, and the black powder continued insoluble.
7. Reduced to powder and very strong nitric acid poured on it there was no effect, but gradually a very gentle effervescence took place, the ore was decomposed and sulphur became visible.
8. A small bit held at the end of a clay-slip in the flame of the lamp it partially melts and glazes the clay-bit around itself. The flame being directed on it by the blow-pipe it melts to a metallic ball and spreads a yellow gloss on the clay. The little metallic button, being separated from the clay-bit and beat on the steel plate, extended to a thin and hot plate which was flexible like lead.
9. The solution No. 7 afforded colorless octahedral crystals.

No. 3093—Black slate.

1. It feels very light.
2. The lens shows particles of mica in it.
3. Before the blow-pipe it takes fire and burns with a flame like coals, but does not melt, leaving a greyish mass of its former shape and volume. This mass is as hard as the slate. The burned bit put into muriatic acid produced a smell of liver of sulphur.
4. Another burned bit at a strong fire melted quickly at the angles to a glossy black matter. It did not stain silver—was not drawn by the magnet. Put on to silver with a drop of muriatic acid it made some small spots on it.
5. Put into pure muriatic acid it effervesced so slowly as to be scarcely visible, and the smaller bits did not fall to powder or soften. Put in powder into muriatic acid the effervescence was more sensible, but I could not not find that the solution reddened sensibly the flame of a candle.
N. B.—This might prove a new test.

No. 3912.—Carbonate of lime. St. Andreasberg.

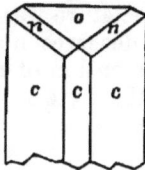

$$oc = 90°$$

$$no = 127° \ 30'$$

$$nc = 142° \ 30'$$

From the above figures it is probable that the faces n are those of the rhombohedron, h, fig 7, Haüy, though the angles differ by 3° 15'.

$$[- \tfrac{1}{3} R : 0 = 127° \ 15'. \quad I.]$$

No. 3926.—Black lead pencil bought at Frankfort. May, 1805.

1. It cost thirty-six kreutzers, or about one shilling and two pence, English.

2. Held in the candle the point does not soften or seem affected.

8. A bit heated at the blow-pipe in the spoon emits a copious white smoke without any sensible smell of sulphur, and the smoke settled as a white powder on bodies. The bit of pencil falls into a coarse scaly powder. This powder looked so like the scaly manganese or iron I suspected its being such ; but melted with saltpetre it consumed and did not impart to it the least bit of green.

A bit of the pencil heated with carbonate of soda did not form visible liver of sulphur, but the solution of the mass stained silver.

No. 3926.—Factitious pencil bought at Frankfort in 1805.

1. A bit exposed at the blow-pipe burns with a flame and emits a copious white smoke. A matter remains which falls to powder under the touch and seems to be plumbago.

No. 5763.—Perhaps Fluorspar, from a lead mine, Matlock bath, in Derbyshire, 1799.

1. Powdered, and put into muriatic acid, there is a momentary effervescence from some particles of carbonate of lime but no sensible diminution of the powder.

2. Heated in sulphuric acid on a bit of glass it effervesced much, but the glass was not depolished.

8. Sulphate of soda formed hydrated sulphate of lime in the solution No. 1.

4. It melted with carbonate of soda, with effervescence, and formed a transparent glass, with opaque white quartz in it which more alkali did not dissolve.

5. This stone scratches glass.

6. The glass (4) was treated with muriatic acid ; the whole did not dissolve.

7. This muriatic solution exhaled dry, left no crystals on adding water. On drying again, and heating more, and adding a small quantity, a dark matter, probably oxide of manganese, was left.

Sulphuric acid added to this solution formed no immediate precipitate, but one of hydrated sulphate of lime formed.

These minute experiments are recorded for a considerable number of specimens. It may be that there were many more of them than have been preserved. They show with what careful and minute accuracy Smithson worked and noted all he did. A large number of these notes were of rocks and clays. This seems to have been the only way in which he busied himself with geology.

A system of chemical nomenclature was made use of in these jottings which, perhaps, deserves notice on account of its curiousness. It is an extension of the astronomical signs, as applied to certain of the metals. They are as follows :

Symbol	Name	Symbol	Name
▽	Water.	⋈	Crystal.
△	Fire.		Precipitate.
∞	Platinum.	⊕	Curdy.
♂	Iron.		Sublimate.
♄	Sulphur.		Baryta.
♀	Copper.		Soda.
☿	Mercury.		Potassa.
O—O	Arsenicum.	♀	Lime.
☉	Gold.	+ ♂	Oxide of Iron.
	Nickel.	O+O	Arsenic.
	Zinc.	+ O—O	Arsenious Acid.
☽	Silver.		Lime Water.
+	Oxygen.	Ô	Magnesia.
⋈	Silica.	+O-O-	Barium Chloride.
△	Carbonic Acid.		Fluor-Calcium.
⋮+⋮	Distilled Vinegar.		Carbonate of Lime.

The following extracts illustrate his manner of thinking:

" Chemistry is yet so new a science, what we know of it bears so small a proportion to what we are ignorant of, our knowledge in every department of it is so incomplete, so broken, consisting so entirely of isolated points thinly scattered like lucid specks on a vast field of darkness, that no researches can be undertaken without producing some facts, leading to some consequences, which extend beyond the boundaries of their immediate object," (p. 26.)

" The only requisite for this operation (crystallization) is a freedom of motion in the masses which tend to unite, which allows them to yield to the impulse which propels them together, and to obey that sort of polarity which occasions them to present to each other the parts adapted to mutual union," (p. 31.)

" I doubt the existence of triple, quadruple, &c., compounds; I believe that *all combination is binary;* that no substance whatever has more than two proximate or true elements," (p. 36.)

" Many persons, from experiencing much difficulty in comprehending the combination together of the earths, have been led to suppose the existence of undiscovered acids in stony crystals. If quartz itself be considered as an acid, to which order of bodies its qualities much more nearly assimilate it, than to the earths, their composition becomes readily intelligible. They will then be neutral salts, silicates, either simple or compound," (p. 46.)

It would be interesting to know if this be the first mention of the acid nature of silica; if so, it should be noticed. This was written in January, 1811:

" A knowledge of the productions of art, and of its operations, is indispensable to the geologist. Bold is the man who undertakes to assign effects to agents with which he has no acquaintance; which he never has beheld in action; to whose indisputable results he is an utter stranger; who engages in the fabrication of a world alike unskilled in the forces and the materials which he employs," (p. 70.)

The following passages would not be lost on certain modern philosophers :

" A want of due conviction that the materials of the globe and the products of the laboratory are the same, that what nature affords spontaneously to men, and what the art of the chemist prepares, differ no ways but in the sources from whence they are derived, has given to the industry of the collector of mineral bodies an erroneous direction," (p. 94.)

" There may be persons who, measuring the importance of the subject by the magnitude of the objects, will cast a supercilious look on this discussion (on intumescence); but the particle and the planet are subject to the same laws; and what is learned upon the one will be known of the other," (p. 101.)

" In the arts of an ancient people much may be seen concerning them; the progress they have made in knowledge of various kinds; their habits; their ideas on many subjects," (p. 101.)

" It is in his knowledge that man has found his greatness and his happiness, the high superiority which he holds over the other animals who inhabit the earth with him, and consequently no ignorance is probably without loss to him no error without evil." (p. 104.)

I have thus attempted to indicate the salient parts of Smithson's scientific achievement. More interesting than the work, however, is the worker. He was eminently an experimenter. All through his papers he is found diligently collecting facts before he proceeds to theorize. This is well shown in his very first paper, that on the so-called Tabasheer. Perhaps the most finished of his papers is that " On a Fibrous Metallic Copper," combining, as it does, an ingenious explanation of a singular phenomenon and subsequent confirmatory experiments.

His style, so clear, so direct, and so exact, is a model for scientific purposes. Of this the extracts above given are good specimens. The paper just referred to, on fibrous copper, and that that on native minium are others.

Of his neatness as a manipulator and skill in devising apparatus I have already spoken.

The papers on "Improvements of Lamps" and an " Improved Method of Making Coffee " show his practical turn.

It is in the last paper but one of the book relative to the "Formation of Kirkdale Cave," that we, perhaps, best of all discover the true fibre of Smithson's mind. The paper was a refutation of the idea of the *Reliquiæ Diluvianæ*, which attempted to refer this cave and some bones found in it to the flood of Genesis. Smithson discusses the subject with the greatest cogency, showing the utter failure of the theory to account for the facts. His argument is of the greatest perspicuity and justness, so correctly does he apprehend every point. This discussion has, of course, lost all its interest at this day, but it had not then, when geology was so imperfectly known. In the last section of this paper the subject is the Deluge, and the effects which must have followed. With real eloquence he shows that, if the secondary limestones were formed during the flood, "embalmed cities, with their monuments " would be found in " every limestone quarry." Such antiquities as these being wholly unknown, he concludes that the removal of the effects of the deluge, like the deluge itself, was due to supernatural causes.

"To a miracle, then," he says, " which swept away all that could recall that day of death, when 'the windows of heaven were opened' upon mankind, must we refer what no natural means are adequate to explain. For this stupenpendous prodigy,

" Like the baseless fabric of a vision,
Left not a wreck behind."

INDEX.

157

www.ingramcontent.com/pod-product-compliance
Lightning Source LLC
Chambersburg PA
CBHW021808190326
41518CB00007B/509